Your AI Ally

Living with Smart Assistants

By
James Hawthorne

Your AI Ally

Living with Smart Assistants

Table of Contents

Introduction

In a world rapidly evolving through technological advancements, artificial intelligence stands at the forefront of transformative change. At the heart of this revolution lie smart assistants, digital companions designed to streamline our interactions with technology and enhance the way we navigate our daily lives. From the bustling corridors of corporate offices to the cozy corners of our homes, these AI-powered aids are gradually becoming indispensable, promising a future where convenience and efficiency are not just desires but inevitable realities.

Smart assistants epitomize the fusion of cutting-edge artificial intelligence with user-friendly interfaces, allowing even those with minimal technical expertise to harness their potential. They represent a significant leap in the evolution of AI, building on decades of research to deliver personalized experiences that cater to individual needs, preferences, and routines. As we stand on the brink of this technological paradigm shift, it's essential to understand how these virtual assistants integrate into the fabric of our daily existence, reshaping how we live, work, and interact with the world around us.

The allure of smart assistants lies in their ability to anticipate our needs and offer solutions before we even voice them. They're capable of not only responding to commands but also learning over time to provide increasingly sophisticated and personalized interactions. This makes them fundamentally different from the technology of yesteryears, positioning them as proactive partners rather than reactive tools. With every interaction, they glean insights, adapt, and evolve, making them an essential asset in navigating the complexities of modern life.

The integration of AI into our daily routines often brings a sense of wonder. Imagine waking up to a gentle nudge by your smart assistant, your favorite morning playlist setting the tone for the day, or your virtual companion reminding you of important tasks as you sip your morning coffee. These scenarios are no longer confined to the realms of science fiction; they're today's reality, transforming how we manage our personal and professional responsibilities. AI assists in automating mundane tasks, allowing us to focus on what truly matters, effectively shaping our productivity and efficiency.

However, with great power comes great responsibility. The pervasiveness of smart assistants raises pertinent questions about privacy, security, and our reliance on technology. As these digital entities become more intertwined with our lives, it's crucial to navigate these concerns thoughtfully. Ensuring that our data remains secure and understanding the implications of sharing personal information are vital to leveraging this technology responsibly. It demands a balance between convenience and caution, a delicate dance between embracing innovation and safeguarding our digital footprints.

Moreover, the rise of smart assistants marks a significant shift in the way we perceive human-computer interaction. It's a testament to our progress, yet it also challenges us to rethink the very essence of what it means to be human in an increasingly mechanized world. As we delegate more responsibilities to AI, we must consider the implications on our cognitive faculties, emotional health, and social dynamics. The technology we create should serve to enhance, not diminish, our inherently human traits of empathy, creativity, and connection.

In this evolving landscape, smart assistants offer immense potential for enhancing accessibility and inclusivity. For individuals with disabilities, they can serve as vital tools, bridging gaps in communication, mobility, and independence. By tailoring experiences and providing customizable solutions, AI systems can significantly improve quality of life, creating opportunities for participation and engagement that were previously challenging or inaccessible.

Looking to the future, the trajectory of smart assistants is bound to be exhilarating. They stand to revolutionize industries, from healthcare and education to finance and entertainment, driving new paradigms of service delivery and customer engagement. With advancements in natural language processing, machine learning, and contextual understanding, these assistants could become indistinguishable from human interactions, blurring the lines between human and machine collaboration.

As we embark on this journey, it's essential to foster a spirit of curiosity, experimentation, and ethical reflection. Smart assistants are not just tools but enablers of a new age of technological possibility. They're catalysts for change, inviting us to reconsider how we interact with our environment, how we perform our roles, and how we define success. The aim of this exploration is not merely to understand the capabilities of these digital allies but to inspire a vision of a world where technology serves as a force for good, enhancing our quality of life while respecting our shared humanity.

Chapter 1:
Understanding Smart Assistants

In the tapestry of modern digital landscapes, smart assistants have emerged as quiet stalwarts, orchestrating a symphony of convenience at our behest. These AI-powered entities, born from an unwavering vision of seamless human-machine interaction, are now embedded in the nucleus of our daily routines, acting as facilitators of efficiency and productivity. Imagine a world where the mundane is automated and intricate schedules are deftly managed, all thanks to incessantly evolving algorithms designed to anticipate our every need. As we stand on the precipice of this technological evolution, it's crucial to understand how these digital confidantes are reshaping the contours of our lives. They are not just tools but partners, capable of intuitively learning from our interactions and adapting to our unique preferences. Navigating the intersection of innovation and practicality, smart assistants promise a future where enhanced personal efficiency is not just a possibility but an expectation, offering a glimpse into how AI can be both the architect and sustainer of enhanced living.

Introduction to Smart Assistants

Smart assistants have rapidly woven themselves into the fabric of our everyday lives, fundamentally altering the way we interact with technology. What began as simple voice-activated systems has now evolved into sophisticated entities capable of performing complex tasks, thanks to the power of artificial intelligence. They're more than just voice-activated gadgets; they're an integration of advanced machine learning algorithms, natural language processing, and vast data management systems that understand, learn, and adapt to user preferences.

The cornerstone of these digital aids is their ability to interpret human language and comprehend context. Unlike their passive predecessors, today's smart assistants can converse with users, providing precise responses to queries and, importantly, translating these into actions. This capability is due to advancements in natural language processing, which allows for an almost intuitive interaction process. By assessing not just the words but also the intent and sentiment behind them, smart assistants transform how users engage with technology, bringing a seamless fluidity that feels less like a command and more like a conversation.

In recent years, the integration of smart assistants into various devices has made them ubiquitous. They're embedded in our smartphones, home automation systems, cars, and a wide range of consumer electronics. This pervasive presence ensures that users have access to their smart assistants wherever and whenever needed, enhancing convenience and accelerating the adoption rate. A simple voice command can now turn on the lights, play your favorite playlist, remind you of meetings, and even control the thermostat – all without lifting a finger. The ease and efficiency with which tasks can be completed through voice commands are truly revolutionary.

For the tech-savvy, smart assistants represent more than just convenience – they symbolize a shift towards a more intelligent, intuitive future. By automating mundane tasks, these digital aides free users to focus on what really matters, be it personal growth, hobbies, or family time. Additionally, for the younger digital natives or the busy professional, these systems form an essential pillar for improving productivity and personal efficiency. Tasks like managing schedules, tracking fitness goals, and even assisting with schoolwork or work presentations are streamlined, shifting the narrative from doing more to achieving more with less effort.

Moreover, the continuous self-improvement mechanism built into these digital aids means they're always learning, evolving to better suit the user's habits and needs. AI algorithms analyze usage patterns to predict what users might require before they even ask. This anticipatory approach enhances user experience, making interactions feel more personal. Imagine waking up, and your assistant has already brewed your coffee, set the room temperature, and listed out your day's agenda based on your previous routines. This level of personalized service is what makes smart assistants so compelling.

In terms of enhancing productivity, smart assistants are powerful allies. They can transform how we manage our work, from streamlining communications to integrating with office software. The idea is simple: by delegating routine decisions and administrative tasks to AI, humans can focus on creative and strategic thinking. The productivity boost here isn't just about doing things faster but doing them smarter. This AI-powered efficiency is creating a paradigm shift in how individuals and organizations view traditional workflows and time management practices.

The integration of smart assistants into education brings another dimension to their utility, offering the potential to democratize learning and make education more accessible. Personalized learning experiences tailored by AI can meet students where they are, fill in learning gaps, and advance their knowledge at a suitable pace. This bespoke approach can cater not just to individual learning speeds but also styles, making education more inclusive. Smart assistants become personalized tutors, offering a blend of instruction, encouragement, and exploration that classroom settings may not always provide.

However, as with any transformative technology, the rise of smart assistants isn't without its challenges. Privacy concerns loom large as these systems collect and process vast amounts of personal data. It's crucial to address these issues head-on, establishing trust through transparent data practices and robust security measures. This transparency will be vital in keeping users engaged and willing to leverage the full potential of smart assistants.

Looking to the future, the ambition is clear: to create smart assistants that aren't merely reactionary tools but proactive partners in enhancing life quality. Imagine a world where they not only perform tasks but also improve the decision-making process by offering insightful analytics and suggesting optimal solutions tailored precisely to user needs. As AI technology continues to advance, the roles that smart assistants play in both work and home environments will inevitably deepen, driven by ongoing innovations in AI research and development. They're not just part of our digital evolution; they're central to it. Understanding their potentialities isn't just useful; it's essential for those eager to thrive in an increasingly automated world.

Evolution of AI in Daily Life

To understand the evolution of AI in daily life, one must first appreciate the incremental yet transformative shifts that artificial intelligence has undergone over recent decades. At its core, AI is the emulation of human intelligence by machines, especially computers. This pursuit of mimicking human cognitive functions, like learning and problem-solving, has evolved from theoretical science into practical technology that inhabits our day-to-day existence.

The timeline of AI development traces back to the mid-20th century, where pioneers like Alan Turing set the groundwork for what would eventually blossom into today's sophisticated technologies. Early on, AI was restricted to the realms of academia and military applications. However, as computational power grew and algorithms became more sophisticated, so did the ambition and scope of AI systems.

By the late 1990s and early 2000s, AI began trickling into consumer technology. Devices like PDAs and basic cell phones introduced some of the first user-facing AI features, such as voice recognition. Yet, these features were often clunky and limited by the hardware constraints of the time. Despite these limitations, they ignited the public's curiosity and set in motion a rush of innovation aimed at integrating AI into everyday devices.

The introduction of smartphones triggered one of the most significant leaps in AI adoption. Suddenly, millions of people had access to portable devices packed with the computational heft necessary for more advanced AI applications. Companies seized this opportunity to embed practical AI features, like virtual keyboards with predictive typing and rudimentary voice assistants into their products.

The real game-changer, however, was the introduction of smart assistants like Apple's Siri, Amazon's Alexa, and Google's Assistant. These platforms marked a paradigm shift in human-computer interaction, allowing users to interact with their devices through natural language. The ability to talk to your phone or smart speaker as if it were a person transformed perceptions about what technology could achieve. It wasn't just about performing tasks; AI began understanding context, preferences, and nuances, making interactions more seamless and human-like.

As AI became more pervasive, its presence extended beyond smartphones and computers, infiltrating other domains of daily life. In the home, AI evolved through smart home devices, integrating with everything from thermostats to lighting systems, allowing for an interconnected ecosystem managed by a central smart assistant. In this way, AI helps optimize energy consumption, enhance security, and increase convenience, fundamentally altering how we manage our living spaces.

The workplace has equally been revolutionized by AI, particularly in terms of enhancing productivity and efficiency. With the ability to automate repetitive tasks, analyze data at unprecedented speeds, and even aid in complex decision-making processes, AI tools have become indispensable. Smart assistants facilitate better collaboration and communication, acting as digital secretaries that manage schedules, reminders, and information flow.

Education and personal development have not been left behind either. AI-driven platforms offer personalized learning experiences, tailoring content and pedagogical strategies to the learner's unique preferences and pace. Such customization was previously unimaginable in traditional learning environments, often limited by resource constraints.

In health and wellness, AI applications monitor vital statistics and provide real-time feedback, encouraging healthier lifestyles and proactive healthcare management. This deployment of AI in wearables enables users to make informed decisions about their health based on data trends and predictive analytics.

Despite these innovations, some challenges remain. Privacy concerns have emerged due to the vast amounts of data AI systems process. Companies have to continuously navigate the delicate balance between offering personalized services and maintaining robust security measures to protect user information.

Looking into the future, the continued evolution of AI in our daily lives will likely hinge on how society addresses these issues. Ethical and societal considerations must be weighed against technological advancement. This balance is essential for fostering trust between AI-empowered systems and the humans they serve.

The journey of AI from a theoretical concept to a cornerstone of daily life reflects the dynamism of human ingenuity. As AI continues to evolve, we must remain vigilant, ensuring its benefits are accessible to everyone while tackling the challenges that accompany such powerful technology.

Chapter 2:
Getting Started with Your AI Assistant

Now that you've delved into the world of smart assistants, it's time to harness their full potential to transform your day-to-day life. Setting up your AI assistant is more than just flipping a switch; it's about crafting an experience that fits like a glove—one that understands and anticipates your needs. Begin by seamlessly integrating your device into your ecosystem, synchronizing it with your routines and gadgets. Personalization emerges as the heartbeat of this process: consider configuring features that align with both your lifestyle and aspirations, whether it's optimizing daily schedules or enhancing entertainment experiences. Embrace the opportunity to mold this digital companion to be proactive in managing tasks, thus unleashing newfound efficiency and convenience. As you embark on this journey, remember that your AI assistant can be a catalyst for not just productivity but also inspiration, bridging the gap between potential and reality in the vibrant tapestry of your ever-connected world.

Setting Up Your Device

Embracing an AI assistant starts with a simple yet essential step: setting up your device. This initial process lays the groundwork for a seamless interaction between you and your assistant, allowing you to transcend the mundane aspects of daily life and delve into a realm where technology and human creativity converge. It's a doorway to a future where tasks that once consumed valuable time now happen effortlessly, thanks to a few spoken words or simple taps.

First, ensure that your device, whether it's a smartphone, tablet, smart speaker, or another form of a smart assistant, is compatible with the AI you intend to use. Compatibility often depends on the operating system and the version you have installed, so it's wise to update your device to the latest version. This not only ensures compatibility but also optimizes your device's performance and security features. The process may involve checking the manufacturer's specifications online or consulting the software documentation accompanying the device.

Network connectivity is paramount for any smart assistant. Ensure your device is connected to a stable Wi-Fi network to facilitate uninterrupted service. A robust internet connection is crucial as it translates your commands into actionable insights by communicating with servers that process vast datasets to deliver results with astonishing intuition. If you experience any issues, consider restarting your router or consulting your internet service provider to troubleshoot the connectivity problems.

Next, look at the installation of the correct app or software provided by the AI assistant's manufacturer. Whether it's Siri, Google Assistant, Alexa, or another brand, follow the installation steps carefully. Most apps guide you through the setup process with instructions that are easy to follow. They will request permissions to access contacts, location, microphone, and other features necessary for optimal functioning. These permissions enable the AI assistant to interact with your environment dynamically and personalize its responses and actions based on contextual information.

Once you have installed the appropriate software, the customization begins, allowing you to personalize your AI experience. This is where you configure voice recognition settings, ensuring that the assistant understands and responds only to your voice, thereby enhancing security and privacy. Voice training might be necessary, which involves repeating certain phrases so the assistant can better recognize your speaking patterns. This process underscores the fusion of human uniqueness with artificial intelligence, tailoring the assistant's functions specifically to you.

Set up user preferences like language options, preferred units of measure, and voice or visual output settings. This allows the assistant to deliver information in a way that's most meaningful to you, enhancing the quality of interaction. This adaptability is what makes smart assistants powerful tools in personal efficiency. It feels as though you are conversing with a device that not only understands your language but also shares in your daily endeavors.

For those who prioritize privacy, it's important to delve into the security settings. Most AI assistants offer customizable options to manage data collection and sharing. By understanding what data is collected and how it is used, you gain control over your digital footprint. Adjusting settings to limit data storage or opting out of certain data-sharing policies can ease concerns about personal information misuse.

Integrating other smart devices with your AI assistant can further enhance its capability. From smart thermostats to lighting systems, each additional device brings new possibilities for automation and control. This interconnectedness, often referred to as the Internet of Things (IoT), enables seamless operation across diverse aspects of home and work environments. Enabling these integrations might require additional apps or skills, which are typically accessible through the assistant's platform.

It's also important to configure emergency settings on your AI device. These can include contacts for emergency calls or sending pre-defined messages in situations where you might be unable to do so manually. Such features demonstrate how smart assistants move beyond mere convenience, playing vital roles in personal safety and emergency preparedness.

Finally, test your setup to ensure that all functionalities are working as expected. Experiment with different commands and routines to explore the full range of capabilities your AI assistant offers. This exploratory phase allows you to discover not only the problems that might need troubleshooting but also the surprising ease with which complex tasks are handled.

In conclusion, setting up your device is more than a mere configuration; it's a transformative process that aligns technology with your unique life rhythms and preferences. With every setting adjusted and each preference noted, your AI assistant becomes not just a tool but a companion in navigating the complexities of modern life. Through careful setup and personalization, you harness the power of artificial intelligence to enhance productivity and simplify everyday challenges, pushing the boundary of what technology can achieve when thoughtfully integrated into our lives.

Customizing Features for Optimal Use

One of the most compelling aspects of AI assistants is their ability to be molded into tools that fit our unique lifestyles. Customizing these features isn't just about setting preferences—it's about transforming your assistant into a personalized digital ally, enhancing your daily routines and interactions.

Let's start with voice recognition, a fundamental feature of any AI assistant. By training your assistant to recognize your voice, you can tap into a world of efficiency and security. This not only minimizes the risk of unauthorized access but also streamlines interaction by instantly identifying you and adapting to your preferences. Imagine walking into your home, and your assistant recognizes your voice, greeting you with your favorite music and turning on the lights just the way you like them.

Next, consider the power of routine automation. Whether you're prepping for a productive day or winding down in the evening, setting up routines allows your assistant to execute a series of actions based on your commands. You might say, "Good morning," triggering your assistant to offer a summary of upcoming meetings, weather forecasts, and even a motivational quote to start your day with the right mindset. By integrating your calendar and task lists, your AI can seamlessly streamline these routines, making them an integral part of your day.

Personalization goes further with contextual understanding. Configure your assistant to grasp nuanced contexts by customizing its location services and setting preferences. This capability turns your AI into a proactive problem solver, offering route suggestions to avoid traffic, adjusting your home's thermostat when you're en route, or suggesting local eateries when you're traveling. Providing timely, location-based information is one key way an AI elevates its utility beyond basic interactions.

Integrating third-party applications further enhances your AI assistant's functionality. By linking services you frequently use—like task managers, note-taking apps, or streaming platforms—your assistant becomes a more effective manager of your digital ecosystem. Enabling this interoperability allows your assistant to provide comprehensive assistance across different platforms, making switching between apps a thing of the past.

Moreover, consider leveraging the power of natural language processing to craft custom commands. This feature grants a level of interaction that mirrors natural conversation. Instead of adhering to rigid command structures, you can set up personalized keywords or phrases that trigger specific actions. This flexibility not only makes your assistant more intuitive but also provides a space for creativity where you can tailor interactions to be as unique as the tasks at hand.

Notification management is another area ripe for customization. Customize how you receive alerts and updates to prevent notification fatigue and ensure you don't miss what's truly important. Whether it's through auditory cues, visual alerts, or silent vibrations, tailoring these settings ensures your assistant communicates with you in a manner that suits your lifestyle, whether you're in a bustling office or a quiet home environment.

For those leaning into the realm of smart home integration, customizing your assistant's interface with various smart home devices can revolutionize everyday living. Imagine controlling your lighting, security systems, and appliances through simple voice commands. Syncing these devices not only enhances convenience but also paves the way for energy savings and a more sustainable living environment, all controlled by the assistant you've tailored to your personal preferences.

Security and privacy settings should not be overlooked when customizing your AI assistant. Ensuring that your assistant operates in a way that maintains your privacy while optimizing performance is critical. Delve into the settings and find the balance between accessibility and security that makes you comfortable. Customizing data sharing and storage preferences can enhance privacy without compromising on functionality.

Lastly, don't forget the aesthetics. Yes, even your AI assistant can have a style! Adjusting the interface to match your personal taste or even the mood you wish to set can add an element of fun to your interactions. Whether it's through changing the assistant's voice, selecting themes, or even setting up custom personas that respond differently based on your requirements, these adjustments make using your assistant a more delightful experience.

In conclusion, customizing your AI assistant for optimal use is about more than tweaking settings—it's about enhancing your interaction with technology to reflect your unique lifestyle. As you delve deeper into these customizations, you'll find that these adjustments not only improve efficiency and convenience but also enrich your everyday experiences, turning potential obstacles into opportunities for growth and productivity. With a personalized AI at your side, the possibilities for managing your digital world are limited only by your imagination.

Chapter 3:
Simplifying Daily Tasks

Harnessing the power of smart assistants, we can transform the mundane into the extraordinary by automating daily tasks and effortlessly managing our schedules and reminders. These AI companions adapt to our routines, learning from patterns to proactively suggest actions that keep our lives running smoothly. Instead of juggling multiple apps or devices, imagine having a singular, cohesive interface that understands your needs and responds in an instant. Forgetting a meeting or anniversary becomes a thing of the past as smart assistants seamlessly sync across calendars and devices, ensuring you're always in the loop. Through simple voice commands or subtle prompts, they guide us through the day, turning chaos into harmony. With this technology integrated into our lives, we not only save time but redefine our relationship with the tools that simplify tasks, allowing us to focus on what truly matters.

Automating Routine Activities

In our fast-paced world, finding ways to offload routine activities not only frees up valuable time but also allows us to channel our energy towards tasks that inspire creativity and innovation. Smart assistants have leaped to the forefront of this technological evolution, offering robust solutions that make automating daily life far more accessible. Whether it's managing your morning routine or ensuring your home is in perfect order while you're away, these digital companions are proving indispensable.

The brilliance of smart assistants lies in their ability to seamlessly execute multiple repetitive tasks without error or fatigue. Consider something as mundane as the morning ritual. With a simple voice command or a series of scheduled actions, you can automate everything from adjusting the thermostat to a comfortable morning temperature, to starting your coffeemaker, and even tuning into your favorite morning news podcast. This isn't just about convenience; it's about crafting a morning experience tailored precisely to your preferences, every single day.

Beyond personalization, smart assistants offer the potential for greater efficiency both at home and beyond it. Think about the minutiae of day-to-day activities: grocery shopping, paying bills, or ordering household essentials. By automating these tasks through AI, you can ensure they're done consistently and promptly without the need for constant human input. Smart assistants can track your supply of everyday items and reorder them when they get low, making sure you're never out of essentials. This not only sidesteps the hassle of last-minute store runs but also optimizes inventory management at home, reducing waste and conserving resources.

Moreover, smart assistants integrate easily with a variety of applications and services, providing a central hub for managing tasks. They can schedule appointments, send reminders, and even synchronize various calendars to prevent overlaps. With additional integration, AI can also analyze your schedule over time, suggesting when to take breaks or how to organize meetings for maximum productivity. By serving as both a personal assistant and a strategic advisor, they keep you organized and efficient, all while requiring minimal input from you.

The office isn't left behind when it comes to automating routine tasks. Smart assistants can streamline workplace efficiency by automating email sorting, setting up conference calls, and preparing meeting agendas. They can even facilitate project management by tracking deadlines, synthesizing reports, or distributing tasks among team members. With these capabilities, AI doesn't replace humans but rather augments their ability to perform at their best.

Security and privacy often accompany discussions on automation. Smart assistants handle vast amounts of personal and operational data, which naturally raises questions about confidentiality. Ensuring that these technologies are secure involves a collaborative effort from developers, policymakers, and users. By promoting stringent security practices and keeping abreast of developments in AI ethics, users can safely harness the power of automation while safeguarding their privacy.

Taking it a step further, smart assistants are beginning to understand the complexities of human behavior, crafting automation solutions that not only perform tasks but do so intelligently. Natural Language Processing (NLP) advancements allow interactions with AI to feel more intuitive, as if instructing a human assistant rather than inputting commands in a system. This subtler form of interaction not only increases user engagement but shifts the relationship from one of utility to collaboration.

Impressively, this context-aware functionality means that smart assistants can adapt over time, learning preferences and refining their performance to suit user habits. They learn when you like things done, how you prefer them to be executed, and even predict when you might want to reorder based on consumption patterns. As it becomes more sophisticated, AI will offer even more granular control, perhaps suggesting automation improvements before you even realize they're necessary.

In industries like healthcare and education, the automation of routine activities is transformative. AI-powered assistants can manage patient records, campaign appointments, and even monitor health conditions remotely with impressive accuracy, all while reducing the administrative burden on professionals. In education, these technologies are beginning to lighten the load of administrative tasks, allowing educators to focus more on teaching and less on paperwork. Through automation, these sectors can offer more personalized experiences without additional strain on human resources.

The transformative potential of automating routine activities with AI is vast. As the technology continues to develop, we can anticipate even broader applications, with increasingly intuitive interfaces and capabilities that will unlock new levels of efficiency and productivity. The true promise of automation lies in its ability to change routine activities into opportunities for growth and its capacity to allow individuals and businesses alike to focus on what truly matters. Embracing this change won't just simplify our daily tasks; it will redefine them, ushering in an era where routine is reimagined through the lens of technological ingenuity.

Managing Schedules and Reminders

The relentless pace of modern life brings with it an intricate web of commitments, deadlines, and to-do lists that often seem insurmountable. Enter AI-powered smart assistants, a technology designed to transform the chaotic into the cogent, enabling individuals to navigate their schedules with newfound ease. Managing schedules and reminders isn't just about avoiding an overdue payment or missing an appointment. It's about paving the path to a more structured and serene day-to-day existence.

For many, the process begins with simply integrating your AI assistant into your routine. It's as easy as issuing a command or typing a short request. Your smart assistant acts as a diligent personal secretary, seamlessly syncing with your calendar apps and email accounts. This integration ensures that all your commitments, be they meetings, family gatherings, or the kids' soccer games, sit neatly in one place. With a few quick updates, your device becomes a comprehensive planner, crossing off a significant chunk of the mental work that often overwhelms us.

One of the most compelling features of a smart assistant is its ability to learn and adapt. By analyzing your routines and preferences, the assistant can anticipate your needs, suggesting optimal times for events and reminders. Whether it's nudging you before a longstanding doctor's appointment or reminding you to call a colleague after wrapping up a meeting, this subtle guidance enhances punctuality and preparedness.

In an age where professional and personal lines blur, the value of setting boundaries cannot be overstressed. Here, AI assists by offering features such as "Do Not Disturb" modes during working hours or family dinners. It's not just about remembering appointments; it's about consciously choosing when to focus and when to switch off. This feature, amalgamated with schedule management, lets you safeguard essential downtime without slipping into the trap of overcommitment.

Setting up reminders through your smart assistant can become a ritual in itself. Some prefer to do this at the start or end of the day, contemplating upcoming tasks and chores while orchestrating their order in the most efficient manner. This practice encourages mindfulness and foresight, cultivating a sense of control over one's schedule. After all, a well-planned day starts with an effectively managed reminder system.

Now, think about how reminders have evolved. From sticky notes and diaries to notifications popping up in our digital spaces, smart assistants now elevate this experience by adopting natural language processing capabilities. This means you can engage with your device as if you were talking to a human, making it intuitive and aligned with your thought processes. You don't need to remember obscure commands; casual conversation does the trick.

Consider the integration of geolocation into schedule management. If location-based reminders are set, your AI assistant will prompt you to pick up dry cleaning only when you pass the area. Such geofencing capabilities are a testament to the technology's prowess in adding value by optimizing time management, reducing redundancy in tasks, and maximizing productivity in often-overlooked ways.

Some might worry that reliance on technology for such personal management could dull their mental acuity or foster dependency. While this is a valid concern, the key lies in balance. Smart assistants are tools designed to augment human capabilities, not replace them. When utilized effectively, they reinforce organizational habits instead of undermining them. The sense of free-flowing control and the elimination of stress associated with forgetting is, in fact, empowering.

Moreover, in professional environments where deadlines are tight and expectations high, the precision and reliability of AI reminders and scheduling can't be underestimated. Executives and employees alike find solace in knowing their day is mapped accurately, with reminders serving as timely nudges rather than disruptive alarms.

The simplicity of inviting or including guests in meetings is another task made easier through voice commands. Your smart assistant can handle sending out calendar invites or scheduling conference calls, allowing you to focus on more strategic or creative tasks. With this kind of assistance, meetings become more than just gatherings; they're now efficiently orchestrated events that respect everyone's time.

As AI continues to evolve, we can expect further enhancements to schedule and reminder management. Imagine assistants that not only learn from individual habits but also aggregate data from various sources to predict broader trends and shifts that might impact your personal schedule. The potential to overhaul traditional models of time management aligns with the greater goal of simplifying life in our digital age.

In conclusion, managing schedules and reminders through AI-powered smart assistants bridges the gap between chaos and control. It offers us a vision of life where technology works alongside us, subtly guiding our daily paths with precision and predictability. Through these devices, we glimpse a future where our energies and attentions can focus more on the substance of our endeavors than the logistics, redefining efficiency and personal productivity for generations to come.

Chapter 4:
Enhancing Productivity at Work

In our increasingly digital workplaces, AI-powered smart assistants are transforming how we approach productivity, allowing us to work smarter, not harder. These intelligent tools streamline communication, breaking down barriers that once slowed collaboration. By integrating seamlessly with office software, they help manage information overload and ensure that every meeting or document is just a voice command away. The real magic, however, lies in how these assistants anticipate our needs, learning and adapting to our work habits to support us more intuitively. The result is a workflow that feels organic and effortless, enabling professionals to focus on creative problem-solving and strategic thinking rather than mundane tasks. As modern work environments evolve, the role of AI becomes not just supportive but essential, forging a future where efficiency is paramount and human potential is maximized.

Tools for Efficient Communication

In the modern workplace, communication is undeniably crucial. As more companies embrace digital transformation, the way teams communicate and collaborate is evolving at an unprecedented pace. Smart assistants, powered by AI, are revolutionizing the way we interact, both in personal and professional contexts. These tools, when properly leveraged, can significantly enhance productivity, streamlining communication processes that once took considerably more time and effort.

One of the primary ways smart assistants boost communication efficiency is through managing emails and messages. Instead of sifting through countless emails or texts, smart assistants can sort, prioritize, and even respond to messages on your behalf. This increases productivity by allowing workers to focus on more complex tasks while maintaining effective communication channels. For instance, AI can automatically categorize incoming messages by urgency, sender, or topic, ensuring that the most important information reaches you first.

Furthermore, scheduling meetings can often become a logistical nightmare, particularly in large organizations. Smart assistants simplify this process by integrating with calendar systems and autonomously coordinating timings, suggesting available slots based on participants' schedules. This reduces back-and-forth communications, making the process more efficient. Some advanced AI tools can even predict optimal meeting times based on past patterns and preferences, avoiding potential clashes and maximizing attendance.

Virtual meetings themselves have been revolutionized by AI. Today's smart assistants can assist in real-time transcription services, ensuring that meeting notes are accurate and comprehensive. This aids participants who might have missed parts of the discussion or those who couldn't attend at all, allowing them to catch up with minimal delay. Moreover, AI-driven language translation tools enable seamless communication across borders, breaking down language barriers and fostering inclusive dialogue in multilingual settings.

Another vital tool in modern communication is voice command interface. Instead of typing inquiries or commands, professionals can use voice-activated assistants to communicate needs swiftly. Whether it's dictating an email, setting a reminder, or searching for information, speaking is generally quicker than typing, increasing the pace at which tasks can be completed. Additionally, with speech recognition technology constantly improving, these tools are becoming more effective and accurate.

Smart assistants also contribute to efficient communication by managing and organizing digital documentation. Digital notes, shared documents, and collaborative platforms are integral to information exchange in workplaces. AI can tag, classify, and archive documents automatically, ensuring that team members can easily retrieve information when needed. With tools like natural language processing, smart assistants can also help summarize documents, extracting key points and making it easier for individuals to digest information quickly.

Integration is key when it comes to maximizing the benefits of these AI tools. Smart assistants typically can sync with other office software, allowing for a cohesive communication system. For example, integrating with project management tools enables seamless information flow between different departments, ensuring everyone stays on the same page. This comprehensive connectivity facilitates better collaboration, reduces misunderstandings, and streamlines workflows.

Security is another pivotal factor in communications. Modern smart assistants come equipped with advanced encryption and authentication features to protect sensitive information. AI tools can automatically manage permissions and access control, granting or restricting data access based on user roles. This level of security ensures that teams can communicate openly without worrying about unauthorized data breaches. Additionally, by monitoring communication patterns, AI can help identify potential threats and notify IT departments to take proactive measures.

Despite the profound advantages, the human element in communication remains irreplaceable. Therefore, it is crucial to use smart assistants as supplementary tools, enhancing rather than superseding personal interaction. Encouraging a harmonious balance between AI and human capabilities nurtures an environment where technology aids communication but doesn't diminish the value of personal input and presence.

Looking to the future, continuous evolutions in AI promise even more advanced communication capabilities. Machine learning algorithms are getting better at understanding context, tone, and intent, making interactions with smart assistants more natural and intuitive. This development might soon enable these tools to suggest more personalized communication strategies, tailored to individual communication styles and preferences.

In conclusion, by embracing AI-powered smart assistants, organizations can unlock a new dimension of productivity in communication. These tools are integral to creating dynamic and efficient workplaces, enabling teams to focus on their core tasks without getting bogged down by communication hurdles. The key to successful implementation lies in strategically integrating these tools into daily workflows, ensuring that they complement human efforts and maintain the essence of interpersonal interaction within professional environments.

Integrating AI with Office Software

In today's digital landscape, office software serves as the backbone of professional productivity. Whether drafting a document, coordinating a project, or managing data, these tools are crucial to work execution. Integrating AI into office software transforms these platforms from static tools into dynamic partners in productivity. AI can automate tasks, provide insightful analysis, and enhance collaborative efforts, ushering in a new era of workplace efficiency.

Imagine sitting at your desk, overwhelmed by the number of tasks you need to complete. You open your word processor, and instead of manually correcting your text, an AI-powered feature suggests edits for clearer and more compelling communication. Simultaneously, your calendar application, enhanced by AI, reviews your schedule, optimizing it to prevent overlaps and even suggesting ideal times for meetings based on the availability of all participants. The mundane becomes manageable with AI, freeing you to focus on strategic tasks that require human creativity and critical thinking.

AI integration is particularly transformative in spreadsheet applications. AI tools can automatically identify patterns in data, generate complicated formulas, and create interactive dashboards with minimal input. This capability not only saves time but also reduces the likelihood of human error, providing accurate data insights that inform better decision-making. For instance, AI-driven models can predict sales trends by analyzing historical data, helping businesses adjust their strategies proactively rather than reactively.

Furthermore, AI's role in office communication tools cannot be overstated. AI chatbots within email platforms streamline communication by sorting, prioritizing, and even responding to messages when appropriate. This reduces the noise of overflowing inboxes, ensuring that the most urgent communications are addressed promptly. Real-time translation services enable seamless communication across global teams, breaking down language barriers and fostering a more inclusive workplace environment.

Document collaboration is another area where AI shines. As teams become more diverse and distributed, AI-powered collaboration tools facilitate smoother workflows. AI can track changes, suggest improvements, and automate version control, preventing the common pitfalls of manual collaboration. Imagine drafting a report with colleagues in different time zones and having AI harmonize the collaborative effort by suggesting edits and even detecting and resolving conflicting versions seamlessly.

The integration of AI into presentation software also revolutionizes how ideas are communicated within organizations. AI tools can generate slide layouts, provide design suggestions, and even offer real-time feedback on presentation effectiveness by analyzing audience engagement. This ensures that communication is not just about the information being conveyed, but also about the most effective means of delivery, maximizing impact.

AI's ability to enhance productivity doesn't stop at individual tools. When AI-oriented platforms are integrated across an organization's entire suite of office software, the effect is synergistic. Imagine the seamless flow of information and tasks from emails to calendars, from documents to spreadsheets, all overseen by AI that anticipates needs and suggests actionable improvements. This holistic integration empowers teams to work smarter, reducing the cognitive load and fostering a culture of efficiency.

However, as with any technological integration, there are challenges and considerations to address. Security and privacy concerns are paramount. When dealing with AI systems that handle sensitive information, it's crucial to implement robust encryption and data management protocols. Additionally, ensuring transparency in AI operations helps build trust among users who may be wary of automated processes.

Training employees to effectively use AI-enhanced tools is also essential. While younger, tech-savvy professionals might adapt quickly, others might require additional support to harness the full potential of these technologies. Offering training sessions and resources can ease this transition, ensuring that all team members benefit equally from the enhanced capabilities.

Looking forward, the integration of AI with office software promises to evolve. As AI technologies become more sophisticated, so too will their ability to understand context, intent, and subtle human interactions. Future iterations might see AI assistants that anticipate needs before they arise, offering solutions that not only address the symptoms of productivity issues but root causes, fundamentally reshaping modern work dynamics.

The potential for AI to rename stale work processes is enormous. It allows professionals to focus on what they do best—create, innovate, and think critically. By automating the routine, minimizing the repetitive, and enhancing the analytical, AI-equipped office software is not just a tool but a catalyst for a more productive and fulfilling work experience.

In essence, AI's integration with office software is a step toward a future where human potential is maximized, liberating us from the drudgery of menial tasks. As we stand on the cusp of this technological revolution, the key is not just to introduce AI into our workflows but to embrace its possibilities and craft a work environment that's as efficient as it is inspiring.

Chapter 5:
Smart Assistants in Education

As educational landscapes continue to evolve with technology, smart assistants are carving out a transformative role, reshaping how students engage with learning materials and educators. These AI-driven tools open up a world of personalized learning experiences, adapting to individual students' needs and learning paces in a way traditional methods simply can't. They act not only as diligent tutors guiding students through complex subjects but also as efficient research assistants, swiftly sifting through vast amounts of information to extract relevant data. This dynamic integration ensures that students spend more time immersed in understanding concepts rather than getting buried in administrative tasks. Beyond academics, smart assistants foster communication between students and teachers, breaking down barriers to create a more inclusive and interactive educational environment. While challenges like digital equity and data privacy need addressing, the promise these smart assistants hold in democratizing education is inspiring. As these digital tools become intricately woven into the fabric of education, their potential to unlock new learning pathways remains vast and exciting.

Personalized Learning Experiences

In the ever-evolving landscape of education, the concept of personalized learning has moved beyond an innovative idea to a practical application, thanks largely to advancements in artificial intelligence. Smart assistants, capable of adapting to individual learning styles and paces, stand at the forefront of this educational transformation. These AI-powered tools are not only reshaping classrooms but also extending educational opportunities into the homes of learners worldwide.

Imagine a learning environment where each student's unique needs are addressed seamlessly. Smart assistants can analyze a student's learning patterns and adjust instructional content accordingly, ensuring a more tailored learning experience. This personalization isn't just about pace; it's about understanding how each student learns best and catering to those preferences. Some may thrive with verbal commands and auditory feedback, while others might prefer visual cues and interactive simulations. These adaptations allow students to engage with material in a way that makes the most sense to them, enhancing comprehension and retention.

The integration of smart assistants in educational settings has opened a world of possibilities for differentiated instruction. Teachers, traditionally tasked with managing a wide spectrum of learning abilities within a single classroom, can now rely on AI tools to diversify their teaching methods. These assistants provide real-time data and recommendations, enabling educators to create more dynamic and inclusive lesson plans. Rather than replacing the teacher's role, smart assistants amplify their capability to reach every student effectively.

Moreover, smart assistants are invaluable in fostering self-directed learning—a skill increasingly necessary in today's fast-paced world. By guiding students through personalized learning pathways, these AI tools encourage learners to take ownership of their educational journey. Students can access resources on demand, receive instant feedback on their progress, and set personalized goals. This model empowers learners, building confidence and fostering a more profound sense of responsibility for their educational outcomes.

The benefits of personalized learning experiences aren't confined to the classroom alone. Homeschooled students, for example, can particularly benefit from smart assistants, gaining access to a spectrum of resources and support that mimic classroom experiences. AI tools can schedule study time, expose students to a diversity of subjects, and even simulate peer discussions through forums and virtual study groups. This support helps alleviate the isolation sometimes associated with homeschooling, enabling learners to engage with a broader community.

With the advancement of natural language processing, smart assistants have become exceptionally adept at understanding and addressing language barriers in education. They can translate content into different languages, making educational materials more accessible to non-native speakers. This capacity to bridge linguistic gaps supports an inclusive learning environment, fostering greater participation among students from diverse cultural backgrounds. In doing so, smart assistants contribute to a more equitable educational landscape.

It's not just students who benefit; teachers themselves gain from this technology. Administrative burdens, such as grading and lesson planning, can be alleviated with AI assistance, allowing educators to focus more on teaching and student interaction. By automating routine tasks, smart assistants free up time for teachers to engage in professional development opportunities, explore new teaching methodologies, and pursue innovative educational projects. This professional growth positively impacts the quality of education delivered in the classroom.

While the potential benefits of smart assistants in education are compelling, it's essential to address the challenges that accompany their implementation. Privacy concerns, for instance, arise as these systems collect and analyze vast amounts of personal data to tailor learning experiences. It's crucial for educational institutions to adopt robust data security measures and transparent policies that protect students' privacy while maximizing the benefits of AI-enhanced learning.

Moreover, the digital divide remains a significant barrier to the widespread adoption of smart assistants in education. Not all students have equal access to the necessary technology and internet connectivity required to utilize these tools effectively. Educational policymakers and technology providers must work collaboratively to bridge this gap, ensuring that all learners can benefit from personalized learning experiences, regardless of their socio-economic status.

Looking to the future, the role of smart assistants in education appears poised for continued growth and innovation. As AI technology advances, these tools will become even more sophisticated, offering deeper insights into student performance and crafting ever more tailored educational experiences. The ultimate goal is a learning environment where every student has the resources and support they need to thrive.

In a world where personalized learning is attainable for everyone, we move closer to an educational paradigm that celebrates diversity in learning styles and embraces equal opportunity for all. The journey toward this reality is ongoing, requiring collaboration, innovation, and a commitment to using technology for the greater good. Smart assistants, with their profound ability to personalize and enhance education, are invaluable allies in this endeavor.

Study and Research Assistance

Amidst the digital revolution in education, smart assistants have emerged as transformative tools in study and research. These AI-powered aids go beyond simple question-and-answer interactions; they become integral partners in the academic journey. Equipped with vast databases and an ability to learn and adapt, they provide invaluable support to learners and researchers alike, helping to streamline processes that once consumed significant amounts of time and cognitive energy.

For students, having a smart assistant is akin to having a personalized tutor available 24/7. The advent of voice-activated technology means that students can now ask questions on the fly and receive swift, articulate responses. This instantaneous support allows them to remain immersed in their study sessions without needing to pause and search through countless resources. Smart assistants help with definitions, conceptual clarifications, and explanations of complex topics, ensuring that students can consolidate their understanding effectively.

Research, by its nature, involves copious amounts of data and literature review. Here, smart assistants come into their own by managing and organizing information seamlessly. They adeptly handle tasks such as finding scholarly articles, comparing study results, and even summarizing papers into digestible insights. Advanced algorithms enable these assistants to filter through layers of academic literature, suggesting relevant articles or data sets that might otherwise remain buried beneath the overwhelming abundance of available information.

Moreover, citation management, a traditionally tedious aspect of research, can be effortlessly handled by smart assistants. They assist researchers in keeping track of sources, ensuring proper formatting according to various citation styles (APA, MLA, etc.), and even alerting them to newer editions or updates in cited works. This not only saves time but also minimizes common errors associated with manual entry, thus enhancing the overall integrity of the research.

Collaboration is a cornerstone of both education and research. Smart assistants bolster this by enhancing communication and coordination among group members. Whether it's scheduling meetings across different time zones or sharing documents and insights, these AI tools streamline the collaborative process. They can even set reminders for important deadlines and milestones, ensuring that projects remain on track and meet the necessary timelines.

Additionally, language translation features enable students and researchers to transcend linguistic barriers, broadening the scope of their collaboration on an international scale. Smart assistants can translate documents and voices in real time, facilitating access to global resources and enhancing cross-cultural communication. This opens endless possibilities for collaborations that enrich personal perspectives and academic contributions.

Accessibility is another area greatly improved by the presence of smart assistants. They cater to users with different learning and accessibility needs. Through natural language processing, they understand and respond to spoken language, making educational content more accessible to individuals with visual impairments or learning difficulties. They can also transform text into speech, creating audio content that aids in comprehension for auditory learners or those with reading challenges.

The customization capabilities of smart assistants allow users to tailor their educational experience. Users can personalize their interaction modes and preferred learning materials. Whether it's setting the pace for studying for an exam or selecting multimedia resources that cater to one's learning style, the flexibility offered helps decrease cognitive load and enhances absorption of information.

In research labs and academic institutions, the integration of smart assistants can optimize routines. Automating administrative tasks like lab inventory management allows researchers more time to focus on experiment and analysis, advancing scientific inquiry with greater efficiency and accuracy. Moreover, the capacity to manage potentially hazardous environments autonomously introduces an additional layer of safety.

Looking towards the future, the role of smart assistants in study and research continues to evolve. With ongoing advancements in machine learning and natural language processing, the potential for even more personalized and engaging educational experiences is substantial. The incorporation of virtual and augmented reality, powered by smart assistants, heralds a new era of immersive learning and interactive research environments.

However, challenges remain. The reliance on these technological aids necessitates careful consideration of data privacy, security, and ethical usage. As AI assistants collect, store, and process vast amounts of personal and sensitive information, stakeholders must ensure policies and practices align with ethical standards and legal requirements to safeguard user data.

In conclusion, smart assistants are revolutionizing the realms of study and research. They empower users by facilitating learning processes, enabling efficient data management, and enhancing collaborative efforts. As technology and AI continue to advance, the integration of smart assistants into education and research will likely deepen, creating dynamic environments that inspire curiosity, innovation, and progress.

Chapter 6:
Health and Wellness with AI

In today's fast-paced world, maintaining health and wellness can feel like a juggling act, but AI-powered smart assistants are stepping up as personal wellness coaches, revolutionizing how we approach our well-being. By seamlessly integrating into our daily lives, these digital helpers offer a custom approach to health management, tracking fitness goals and heart rate fluctuations, even reminding us to hydrate or meditate. The magic lies in their ability to synthesize personal health data and provide actionable insights. Imagine an assistant that not only monitors your health statistics but motivates you with timely encouragement, adapts advice based on real-time data, and suggests personalized routines to meet your unique fitness objectives. This proactive engagement transforms wellness from a task into a seamless part of life, potentially reducing the risk of chronic illnesses and enhancing overall quality of life. As we harness the power of AI, the possibilities for improved health outcomes are not just promising—they're already here, reshaping the narrative of personal health and wellness to make it more accessible and tailored to our individual needs.

Tracking Fitness Goals

In the realm of health and wellness, the fusion of artificial intelligence with fitness is reshaping how individuals set, monitor, and achieve their fitness goals. In an era where data is gold, smart assistants emerge as invaluable allies, providing personalized insights and motivation that were once the domain of personal trainers and fitness experts. With a kind of intelligent foresight, AI doesn't just make suggestions—it becomes an ever-watchful guardian of your fitness destiny.

From the outset, AI can assist in establishing a realistic starting point for your fitness journey. By connecting to wearable technology—such as fitness bands, smartwatches, and even smart clothing—AI can analyze vast amounts of personal data, including heart rate, steps taken, sleep patterns, and more. This data is then contextualized, allowing your AI to generate fitness goals that are not only aspirational but tailored to your own body and lifestyle.

The real power of AI lies in its ability to process data in real-time and adjust fitness plans dynamically. Imagine this: after assessing your current physical activity and dietary habits, your smart assistant can suggest a comprehensive workout regimen that aligns with your lifestyle. If one week you're more sedentary than usual, perhaps due to work demands, your AI might dynamically lower your activity targets for that period, reducing the risk of burnout while keeping you on track toward long-term goals.

AI's contribution extends beyond setting goals; it plays an essential role in keeping you accountable. Regular notifications, progress updates, and gentle nudges keep you on course and can even motivate you when your enthusiasm wanes. Some smart assistants incorporate gamification, turning fitness tracking into a rewarding experience. As you reach milestones, celebrations in the form of virtual rewards or badges fuel your motivation, making the pursuit of health both engaging and enjoyable.

Moreover, the incorporation of social elements into AI fitness programs can foster community and support. Through AI, you can connect with friends or like-minded enthusiasts to share progress, engage in challenges, or simply seek encouragement. Whether it's by sharing leaderboards or virtual high-fives, these interactions can be vital in maintaining motivation and ensuring that fitness remains a fun and social endeavor.

AI also ensures variety in achieving fitness goals. It analyzes trends and patterns from an ever-expanding body of knowledge, tailoring workout routines that evolve with you. Whether it's introducing new exercises to prevent monotony or adjusting routines to avoid injury, smart assistants are able to keep you engaged and safe. This adaptability also encompasses dietary planning, as AI can suggest meal plans or recipes that are nutritious and inclined toward your specific fitness goals.

On the horizon, we see potential advancements transforming AI's role in fitness even further. As AI becomes more emotionally intelligent, it might provide motivational boosts on days when your mood is low. Imagine a digital companion that recognizes not just your physical exhaustion but also your emotional state, offering encouragement or suggesting meditation when needed. This kind of empathetic technology could revolutionize mental wellness in tandem with physical fitness.

Furthermore, as we move toward more immersive technology, augmented and virtual reality could become vital in AI-driven fitness. Imagine exercising in a virtual setting that mimics what you enjoy—be it jogging through a forest or cycling along a serene lakeside—all synchronized through AI to optimize each session according to your performance metrics.

Even with these advancements, it's crucial to remember the balance between using AI and personal judgment. AI provides data-driven insights, but the insights should be combined with personal intuition and bodily awareness. After all, the ultimate goal is not just to reach a fitness target but to embrace a healthier, more balanced lifestyle.

As we navigate the future, AI in tracking fitness goals is poised to emerge as more than just a tool—it's a personalized partner that empowers you to transcend boundaries, helping transform aspirations into tangible achievements. The journey to health becomes less daunting and more attainable through the guiding hand of technology that is forever learning, adapting, and evolving alongside us.

Monitoring Health Data

In the rapidly evolving landscape of health and wellness, AI-powered smart assistants are revolutionizing how individuals monitor and optimize their well-being. As technology continues to advance, the integration of AI in healthcare is empowering users to play an active role in managing their health data. This isn't just about numbers and statistics; it's a means to create a comprehensive picture of one's health and inform better decision-making.

Understanding the importance of health data monitoring begins with recognizing the diversity of data points available. Today, smart assistants can track a wide array of health metrics, from vital signs such as heart rate and blood pressure to lifestyle factors like sleep quality and physical activity. This wealth of data provides insights that were once the domain of healthcare professionals, putting powerful tools in the hands of individuals.

The aggregation and analysis of health data by AI assistants allow for personalized insights that can guide users in making informed health decisions. For instance, patterns in sleep data, when analyzed, could show correlations with periods of high stress, resulting in targeted recommendations for stress management techniques or sleep improvement strategies. By detecting patterns and anomalies, AI can also prompt timely medical check-ups or lifestyle adjustments, potentially catching health issues before they become serious.

Moreover, the ability to continuously monitor health data can inspire positive behavior changes. By providing real-time feedback and reminders, smart assistants encourage users to stay active, maintain hydration, or take prescribed medications on schedule. These small nudges, facilitated by AI, help build healthier habits over time, fundamentally changing the approach to daily health management.

Privacy and data security are paramount in the context of health data monitoring. Users are increasingly concerned about how their delicate personal information is handled. AI systems must be designed with robust security protocols to ensure that data is encrypted and stored safely. Transparency in data management practices is crucial for maintaining trust. Users need to have control over their data, with the ability to decide what is shared and with whom.

Another exciting frontier is the use of AI to predict potential health risks by analyzing trends over time. Predictive analytics within these platforms can identify specific risk factors for developing conditions like diabetes or hypertension, enabling proactive measures to mitigate these risks. Such insights can transform healthcare from being reactive to proactive, ultimately reducing the burden on healthcare systems.

With AI's ability to process and learn from vast datasets, there's potential to tailor individualized wellness plans. By understanding user preferences and adapting to feedback, AI assistants can suggest workout plans, propose dietary adjustments, or recommend stress-relief activities that cater to individual needs and circumstances. These personalized plans are not only more effective but also more engaging, motivating continued use and adherence.

And it's not just individuals benefiting from these advancements. Healthcare providers can leverage AI-generated data to enhance patient care. By having access to a patient's longitudinal health data, doctors can make better-informed clinical decisions. AI can also aid in the efficiency of healthcare systems by streamlining administrative processes, allowing healthcare professionals to focus more on direct patient care.

Smart assistants, by bridging the gap between users and healthcare providers, foster a collaborative approach to health management. This enhances the continuity of care, as both parties are equipped with a more comprehensive and up-to-date picture of the patient's health status. This collaboration enriches the patient experience and leads to improved health outcomes.

Looking to the future, we can anticipate even deeper integration of AI in health monitoring. Emerging technologies like wearable biometric sensors and advanced machine learning algorithms will further enhance the capabilities of smart assistants. Innovations in natural language processing could improve how individuals interact with these systems, making health management even more intuitive and accessible.

However, while the potential is vast, it is crucial to recognize the ethical considerations surrounding AI in health monitoring. This includes ensuring equitable access to these technologies and addressing any biases inherent in AI algorithms. Inclusive design and careful regulation can help ensure that advancements in AI healthcare technologies benefit all segments of society.

Ultimately, the promise of AI in health monitoring lies in its ability to transform healthcare from being episodic to continuous, creating a seamless integration of wellness management into daily life. By empowering individuals with the tools to understand and act on their health data, smart assistants foster a sense of agency and control, leading to healthier and more informed lives.

Chapter 7:
Managing Finances with
Smart Assistants

As modern finance becomes more and more complex, smart assistants step in as powerful allies in our budgeting and financial planning journey. They transform the way individuals monitor their spending, offering seamless expense tracking through real-time notifications and insights. These virtual companions can analyze spending patterns, providing tailored advice that's easy to digest, whether it's about cutting unnecessary costs or finding the best savings account. As they integrate with banks and financial apps, smart assistants not only facilitate smooth financial management but also bring personalized investment insights right to the user's fingertips. This creates an environment where informed decisions are the norm, turning complex data into actionable knowledge that empowers users to achieve their financial goals with ease and foresight. In an era where time is money, these AI tools offer unparalleled efficiency, making financial management an intuitive and less daunting task.

Budgeting and Expense Tracking

Managing personal finances can be challenging in today's fast-paced world. Fortunately, smart assistants are stepping in to redefine how individuals track expenses and manage budgets. With AI integration, these digital helpers offer a seamless experience in financial management, streamlining processes that were once tedious and time-consuming. They analyze spending habits, provide actionable insights, and help build a more disciplined approach to managing money.

Smart assistants can assist with budgeting by categorizing expenses automatically. Using machine learning algorithms, they accurately identify recurring bills, groceries, entertainment, and more. This categorization is often done in real-time, allowing users to see a detailed breakdown of their spending with just a voice command. When individuals understand where their money is going, they can make more informed decisions about their budgets.

Expense tracking has never been easier. By linking financial accounts to a smart assistant, users can receive updates on their daily spending limits, ensuring they stay within their budget. Imagine asking your smart assistant about your latest credit card transactions or current bank balance, and receiving an immediate response. This real-time insight prevents overspending and helps keep financial goals on track.

Moreover, these AI assistants can provide personalized budgeting tips based on an individual's financial situation. They analyze patterns over time and suggest areas for improvement. Whether it's cutting back on coffee shop visits or reallocating funds to a savings account, smart assistants offer tailored advice to bolster financial health. These insights empower individuals to make smarter decisions with their money.

For individuals with specific financial goals, such as saving for a vacation or a new car, smart assistants can be invaluable allies. Users can set financial targets, and the assistant will track progress, offering regular updates and encouragement. This feature is especially motivational, as seeing tangible progress can boost commitment to reaching financial milestones.

Smart assistants also excel in reminding users about upcoming payments. Gone are the days of missing due dates or incurring late fees. By setting up payment reminders, either through notifications or calendar events, these assistants ensure users pay bills on time. This not only helps maintain good credit but also reduces financial stress.

Budgeting applications integrated with smart assistants extend the utility of AI in financial management. Apps like Mint or YNAB (You Need A Budget) can sync seamlessly for a more comprehensive view of financial health. Smart assistants facilitate this integration, encouraging users to engage more with their finances and offering an opportunity for greater control over their monetary lifestyle.

Privacy concerns are often paramount in financial dealings, and smart assistant developers are highly aware of this. They've implemented stringent data protection protocols to ensure users' financial information remains secure. It's vital to understand how these systems protect data, which fosters trust and encourages more widespread adoption of such technology for personal finance management.

The question arises: how do these assistants adapt to changes in lifestyle? Perhaps an individual starts a new job, moves to a different city, or experiences a shift in financial priorities. Smart assistants are designed to evolve alongside their users. Through continuous learning, they adjust recommendations and budgeting tips to align with new circumstances, showcasing the incredible potential for personalized financial management.

The broader impact of smart assistants in finance goes beyond individual benefits. By promoting better budgeting habits, these tools contribute to a more financially literate society. As more people become adept at managing their finances through technology, we could see a ripple effect of reduced financial hardships and increased prosperity at a societal level.

Imagining the future, the potential for smart assistants to revolutionize personal finance seems limitless. Innovations such as predictive analytics could anticipate financial needs before they arise, offering solutions proactively. The marriage of AI and finance is poised to evolve continuously, making budgeting and expense tracking even more intuitive and accessible for everyone.

Despite all their capabilities, smart assistants are not a cure-all solution. Users must engage actively with their financial data to make the most of these tools. A smart assistant can provide information and suggestions, but it's up to the individual to take action. By fostering a disciplined approach and embracing the technology available, individuals can leverage their smart assistants to achieve financial stability and peace of mind.

Investment and Savings Advice

As we continue exploring how smart assistants can redefine personal finance management, their capabilities extend beyond just monitoring expenses and setting budgets. Imagine having a financial advisor who operates around the clock, offering tips and insights precisely tailored to your investment and savings goals. This isn't just a distant fantasy— it's now a reality made possible through the use of AI-powered smart assistants.

Today's smart assistants are equipped with algorithms designed to analyze financial markets and trends. They can alert users to shifts in market conditions, provide updates on stock performance, and even suggest diversifying assets based on current market behaviors. This analysis can be essential for someone navigating the complexities of investing, helping users make informed decisions without needing to become finance experts themselves. The efficiency of these tools lies in their ability to process vast amounts of data quickly, distilling it into actionable insights.

For those looking to venture into investment landscapes, smart assistants offer a range of recommendations from conservative savings accounts to higher-risk growth stocks. Depending on the user's risk tolerance and investment objectives, these AI systems compile and evaluate relevant data from varied sources, generating a customized financial roadmap. This approach not only simplifies the decision-making process but also empowers users to take control of their financial futures.

Furthermore, let's not underestimate the power of these assistants in facilitating savings strategies. Often, building a sound savings plan requires relentless discipline and careful tracking of expenses versus income. Smart assistants step in by regularly monitoring spending habits, offering alternative suggestions to curb unnecessary expenditures, or proposing savings goals that align with future objectives, whether it's purchasing a home or planning for retirement.

Imagine setting a target retirement age with your assistant. It would constantly provide updates on whether you're on track, suggest adjustments to your savings strategy, or point out extra contributions during periods of increased income. By furnishing a dynamic plan, smart assistants ensure that financial progression adjusts to life changes and keeps you aligned with your aspirations.

In the realm of retirement planning, AI's ability to predict economic changes over the long term is particularly advantageous. Smart assistants could simulate potential outcomes based on various economic scenarios, advising on protective measures to safeguard investments. By focusing on long-term growth and potential risks, AI-driven recommendations can help mitigate unforeseen impacts on retirement savings.

Moreover, leveraging AI for investment and savings advice can also encompass tax optimization strategies—an often-overlooked aspect of personal finance management. Smart assistants can prompt users about tax-efficient savings accounts, retirement contributions limits, or possible deductions. They consolidate financial data over the year, aligning this with applicable regulations, resulting in automated, valuable end-of-year financial insights.

Beyond personalized advice, these assistants facilitate seamless integration with financial applications and services. A holistic view of one's financial ecosystem becomes accessible, fostering an atmosphere where synchronized financial management is the norm rather than the exception. This integration is vital for individuals with diversified portfolios spread across multiple institutions.

AI-driven financial planning encourages accountability. Imagine an assistant that regularly checks in with you, reminding you of your goals and offering kudos when targets are met or suggestions when you're veering off course. This positive reinforcement can be a significant motivator in maintaining a disciplined savings and investment routine.

However, as with any tool grounded in data analysis, the accuracy of suggestions by AI assistants relies heavily on the quality and breadth of data available to them. Undoubtedly, the question of privacy arises—how much should one divulge to ensure optimal functionality? Users must strike a balance between sharing information and safeguarding personal privacy, a nuanced dialogue that must be guided by newly emerging digital ethics standards.

Though AI-managed investments and savings might seem novel, they're part of a burgeoning ecosystem set to flourish further. As smart assistants become more adept in understanding user sentiments and behaviors, their predictive capabilities will only enhance, bringing more refined and humanized advice to the forefront. This evolution promises a future where financial independence and literacy are within reach for many more people.

While the financial landscapes are ever-evolving, it's vital to remember that smart assistants are tools to assist—augmented extensions of your strategic thinking, not substitutes for genuine human judgment and discretion. Integrating AI into your financial planning arsenal means wielding a powerful, adaptive ally on the path to economic security and success.

In conclusion, the marriage of artificial intelligence and personal finance provides a compelling case for enhancing one's investment and savings strategies. With an AI assistant by your side, navigating these often-complex realms becomes not just manageable but also empowering. Engaging with these tools transforms how you interact with your finances—improving clarity, driving informed decisions, and ultimately, optimizing for the prosperity of the individual.

Chapter 8:
AI in Home Automation

In the realm of home automation, AI is transforming houses into intelligent ecosystems where convenience, security, and efficiency converge seamlessly. Imagine a world where your voice commands effortlessly adjust lighting, regulate temperature, and control appliances, all while enhancing your home's security features with AI-driven surveillance systems. With smart assistants at the core, these innovations not only streamline daily routines but also foster an environment tailored to your personal comfort. As AI continues to evolve, we're on the cusp of even more sophisticated interactions, where anticipatory systems learn from our habits and preferences, proactively managing our living spaces with minimal input. The future of home automation holds the promise of unparalleled integration, where technology intuitively supports our lifestyles, making our homes smarter, more responsive, and ultimately, more human-centric.

Voice-Controlled Devices

Voice-controlled devices are more than just gadgets—they're revolutionizing how we interact with technology at home. As artificial intelligence (AI) continues to evolve, these devices are taking center stage, enabling seamless interactions and automating everyday tasks. Imagine walking into a room and having the lights adjust to your mood, the temperature set just right, and your favorite playlist start to play—all with just a few spoken words. This is the magic of voice-controlled devices, bringing an unprecedented level of convenience and personalization.

At the heart of these devices are smart assistants, powered by AI, such as Amazon's Alexa, Google's Assistant, and Apple's Siri. They listen intently, understand voice commands, and perform actions by interpreting natural language. This technology transforms how users engage with their homes, making it possible to control appliances, manage schedules, and access information without lifting a finger. The interaction is so natural that it often feels like conversing with a knowledgeable companion who's always ready to help.

One of the most significant advantages of using voice-controlled devices is their potential to enhance daily routines. By enabling hands-free operation, they allow people to multitask more efficiently. For instance, cooking a new recipe becomes a breeze when you can ask your voice assistant to set a timer, convert measurements, or even call a friend for advice—all while your hands are full. The ease of managing these tasks by voice reduces friction and enhances productivity.

The integration of voice-controlled devices with other smart home technologies amplifies their utility. Today, many homes equipped with smart thermostats, lighting systems, and security cameras can be controlled via voice commands. This interconnectedness not only simplifies user experience but also optimizes energy efficiency. For example, you can instruct your system to turn down the heat when you leave, or lock doors and arm security systems from afar, all done using a simple voice command.

However, as these devices become increasingly ubiquitous, they spark discussions about privacy and data security. They continuously process spoken input, which raises legitimate concerns about whether conversations are being recorded or misappropriated. Manufacturers have a responsibility to safeguard user data, and many devices come with privacy features such as mute buttons and options to delete voice recordings. Understanding these features can empower users to enjoy the benefits of voice-controlled devices while maintaining control over their personal information.

The adaptability of voice-controlled devices extends to accessibility, offering tremendous benefits to users with mobility challenges or visual impairments. With voice commands, tasks that were once challenging or impossible become achievable. Adjusting settings in the home, getting real-time news updates, and managing personal schedules contribute to a sense of independence for these individuals. This democratization of technology breaks barriers, making homes smarter and more inclusive.

As AI's capabilities grow, voice-controlled devices are expected to become even more intuitive. With advancements in machine learning, these assistants will not just follow commands, but anticipate needs and make recommendations based on user habits. The future holds the promise of assistants that can discern mood changes through vocal tones and suggest activities or media that align with emotional states, offering a personalized touch that's currently just on the horizon.

The journey of voice-controlled devices is just beginning. As technologists push the boundaries of what's possible, the fusion of voice recognition, AI, and home automation will continue to redefine convenience. This transformative power lies not only in what these devices can do today but in their potential to anticipate, learn, and evolve alongside their users. The ultimate goal is to create environments that respond to us intuitively, making life not only simpler but significantly more enriching.

In a world where time is a premium commodity, voice-controlled devices offer a portal to a more efficient lifestyle. They encapsulate the spirit of innovation by making everyday tasks effortless through the sheer power of spoken word. As we stand at the cusp of further technological breakthroughs, the anticipation for what will come next invigorates the imagination, promising endless possibilities in the realm of AI-driven home automation.

Enhancing Home Security

Imagine returning home to find your property has seamlessly transformed into a fortress of security, all orchestrated by invisible lines of AI code, working tirelessly to ensure your peace of mind. It's a vision that's increasingly becoming a reality as artificial intelligence harmonizes with home automation to redefine domestic safety. The integration of AI in home security systems embodies a profound shift, moving from passive deterrence to active prevention.

The days when a basic lock and key were sufficient to secure our homes are long gone. Today, as security threats evolve and adapt, so must our defenses. AI-powered home security systems now act not just as silent sentinels but as proactive guardians. They learn, anticipate, and respond, offering capabilities beyond the traditional scope of security measures. Smart cameras, equipped with facial recognition, can differentiate between a family member arriving home late and an uninvited guest lurking by the door. In mere seconds, they dispatch alerts to your devices, allowing for quick action.

At the core of AI-enhanced home security is a web of interconnected devices that communicate seamlessly. This IoT (Internet of Things) framework ensures that lights, cameras, alarms, and locks collaborate efficiently, weaving an impenetrable safety net. When an anomaly is detected by a smart sensor, it doesn't act in isolation. Instead, it prompts a cascade of responses across the system—from locking doors to streaming live camera feeds to your smartphone, every action is orchestrated swiftly.

li

Smart cameras with real-time alerts

AI-based face and object recognition

Remote monitoring via mobile apps

Yet, the functionality extends beyond mere deterrence and alert mechanisms. AI brings an inherent capability to learn from patterns. It analyzes vast amounts of data accrued over time—movement patterns around your house or typical entry and exit times—to identify what is usual and what is not. Such a level of understanding equips the system to trigger alarms only when genuine threats are perceived, greatly reducing false alarms.

Moreover, AI enables more adaptive and personalized security protocols. By assessing user-specific patterns and preferences, it tailors its responses to align with your lifestyle. Take, for instance, a scenario where you're expecting a delivery. Your smart assistant can construct temporary access permissions for delivery personnel, ensuring seamless package drop-offs without compromising on security. This nuanced response replaces the typical all-or-nothing approach with a more intelligent solution.

Beyond its immediate functionalities, the influence of AI in home security extends to a broader societal level. Neighborhoods wired with AI-enhanced systems contribute to a larger network of communal safety. Shared alerts and data between households build an environment of mutual vigilance, where information flows both horizontally among peers and vertically to local authorities when needed, fostering a cooperative crime prevention strategy.

Moreover, AI-driven security systems bring an added layer of convenience to everyday life. They provide homeowners with a unique peace of mind that doesn't require constant manual checks or oversight. A single voice command or a quick glance at a mobile app provides real-time updates on the status of your home, allowing you to engage with your security system effortlessly, whether you're relaxing in the backyard or halfway across the globe.

To ensure these sophisticated systems serve their purpose effectively, continuous advancements and updates are crucial. Developers are persistently collaborating to fortify these systems against potential tampering and hacking attempts. This constant evolution not only enhances security but also addresses privacy concerns, ensuring that the trade-off between safety and personal data integrity is minimized.

In the journey of enhancing home security with AI, another crucial aspect is energy efficiency and sustainability. As the environment becomes an increasing focus of global initiatives, AI technology in home security takes center stage to play its part. Devices powered by AI can optimize their operational efficiency by activating only when necessary, thus saving energy and reducing carbon footprints.

Envisioning the future, AI's potential in home security appears limitless, with rapid advancements shaping an era where predictive intelligence and automated decision-making become hallmarks of modern security systems. Whether through biometric entry systems, advanced encryption protocols, or even integrating with broader smart city infrastructures, AI promises a future where home security isn't just about protecting valuables but securing a peace of mind that resonates through every aspect of modern living.

As AI continues to contribute to the evolution of home security, it challenges us to rethink traditional paradigms and embrace a technology-driven approach to safety. No longer confined to manual controls or reactive measures, security today is becoming a proactive experience, intertwined with the rhythm of daily life and seamlessly integrated into the broader landscape of smart home automation.

Chapter 9:
Travel Planning and Management

In today's fast-paced world, the ability to plan and manage travel efficiently can be a major game changer. With AI-powered smart assistants, you can effortlessly book flights and accommodations, all while keeping track of your itinerary with pinpoint precision. Imagine having real-time updates about potential travel disruptions or changes; this is no longer a luxury but a readily accessible feature. These assistants can scout for the best deals and even suggest hidden gems at your destination, turning every trip into a seamless, personalized adventure. As we delve deeper into how these AI assistants can simplify travel planning, it becomes evident that they are invaluable for anyone looking to harness technology for smoother, more enjoyable travel experiences. This leap in travel management showcases the profound potential of AI to transform even the most mundane activities into tasks imbued with efficiency and foresight.

Booking and Itinerary Management

In today's fast-paced world, the concept of travel has transformed from a luxury to an essential component of modern life. Whether it's for business, leisure, or family commitments, efficient travel planning can make a considerable difference. Enter the realm of AI-powered smart assistants, which have begun to revolutionize the way travelers approach booking and itinerary management. Seamlessly orchestrating the multifaceted elements of travel, these digital aides promise not only to save time but also to enhance the experience by catering to individual preferences.

At the heart of booking and itinerary management lies the process of making informed decisions. Smart assistants equipped with AI can sift through vast repositories of data to compare flights, hotels, and transportation options, ensuring that users get the best value for their money. With personalized algorithms that learn from past behavior, these assistants can tailor suggestions that align with the user's travel history and preferences, offering a level of customization that traditional methods could never achieve. This intelligence cuts through the noise and presents options that are meaningful and relevant.

Gone are the days of juggling multiple tabs and applications just to organize a simple trip; now, one can simply voice a command or type a request, allowing the smart assistant to handle the heavy lifting. By understanding natural language, these assistants can interpret complex queries—such as booking a direct flight with specific airlines, choosing a hotel with five-star reviews within a certain distance from the city center, or even arranging ground transportation upon arrival. The ease of use is akin to having a personal travel agent who patiently attends to every detail, day or night.

But it doesn't stop at booking. A significant strength of AI-powered smart assistants is their ability to consolidate travel itineraries into a single, comprehensive platform. They automatically integrate flight details, accommodation bookings, car rentals, and even restaurant reservations into one accessible itinerary. This integration is especially useful for frequent travelers who need to track multiple trips simultaneously. With real-time updates, any changes—such as flight delays or gate alterations—are instantly communicated to the user, providing peace of mind and reducing the stress that often accompanies travel.

There's also the element of interconnectivity with other smart devices and platforms. For example, linking calendar apps to smart assistants ensures that any changes in the itinerary are reflected across all digital devices, keeping users up to date effortlessly. This synchronization is crucial for travelers who manage tight schedules and need their itinerary to be as fluid as their lifestyle demands.

Furthermore, smart assistants can proactively suggest enhancements to trips, tapping into vast networks of user reviews and local insights. Whether it's recommending a popular local eatery, suggesting activities aligned with personal interests, or highlighting cultural events happening during the stay, the goal is to enrich the travel experience beyond the basics of getting from point A to B. This capability turns a standard trip into a curated journey.

The integration of AI in travel management also raises considerations around trust and data privacy. With smart assistants accessing sensitive information like passport details and credit card numbers, it's imperative for service providers to employ stringent security measures. Users must feel confident that their data is handled responsibly and that their privacy is maintained at all times. AI developers continually work on enhancing encryption and other security protocols to ensure these technologies remain not only convenient but also safe.

In cases where travel plans change—a common occurrence—smart assistants offer the flexibility to adjust bookings with minimal hassle. By understanding cancellation policies, alternative travel routes, and optimal rebooking strategies, these digital companions offer solutions that keep the user's best interests at heart. This agility is particularly valuable when dealing with unforeseen disruptions, as the assistant can quickly identify and propose alternatives, maintaining the flow of the journey.

Among the many benefits of AI-driven itinerary management, personalization remains a standout feature. By analyzing user behavior and preferences, smart assistants fine-tune travel suggestions to match individual tastes and needs. This personalization helps build a more enjoyable and less stressful travel experience, transforming routine business trips into opportunities for personal enrichment and exploration.

Looking to the future, the potential for even more sophisticated booking and itinerary management solutions looms large. As AI technology advances, the depth and breadth of features available to travelers will expand. We can anticipate even greater predictive accuracy, enhanced user interfaces, and deeper integration with emerging technologies such as augmented reality and the Internet of Things (IoT). This evolution promises a paradigm shift in how we conceptualize and experience travel.

Ultimately, the intersection of AI-driven smart assistants and travel itinerary management signifies much more than a convenience; it's a gateway to new possibilities. By freeing users from the mundane aspects of travel planning, AI enables a focus on what truly enriches the travel experience: the joy of discovery and the connections we make along the way. As we continue to embrace these technologies, the prospect of a more efficient, personalized, and exciting world of travel is within reach.

Real-Time Updates and Adjustments

Travel planning has evolved from flipping through thick tour guides to leveraging the power of AI for seamless and spontaneous adjustments on the go. Today's smart assistants offer real-time updates, transforming rigorous planning sessions into dynamic journeys filled with spontaneity and efficiency. Picture this: as you're about to board a train, your smart assistant alerts you to a delay, providing alternative routes or suggestion to catch a nap at a nearby lounge. This kind of timely information is a game changer for travelers who crave adaptability and reliable insights.

Smart assistants integrate with a variety of real-time data sources to deliver up-to-the-minute updates. They pull information from airlines, weather services, public transportation networks, and even traffic cameras to provide the most current information possible. If a flight gets canceled or a storm moves in, your smart assistant can swiftly notify you and offer assistance in rebooking flights, finding accommodations, or altering itineraries. This means less stress and more time enjoying the journey itself.

Moreover, these AI systems can forecast potential issues before they disrupt your plans. Thanks to machine learning algorithms and vast data analysis capabilities, smart assistants don't just react to changes—they can often anticipate them. By studying patterns and historical data, they might warn you of a usual traffic jam or recommend an earlier departure time for a smoother travel experience. It's about making travel smarter, not harder.

Consider the urban adventurer navigating through a new city. With a smart assistant in your pocket, you're no longer bound to static maps or preset schedules. You can toggle between preferred modes of transportation—whether it's a scooter, ride-share, or subway—based on real-time traffic conditions and your own preferences at that moment. You have the option to pursue dining recommendations or landmark visits, which change dynamically depending on your current location and time constraints. Your travel experience is as flexible as you need it to be.

As we delve deeper into the capabilities of real-time updates offered by AI assistants, the element of personalization cannot be understated. What sets these assistants apart is their ability to tailor suggestions and updates that suit your travel style and preferences.

Whether you're a backpacker on a budget or a business traveler with specific needs, these AI companions learn your patterns and adapt their recommendations accordingly. This isn't just about efficiency; it's about experiencing travel your way.

Incorporating real-time travel updates isn't merely practical—it's also a safeguard. Safety is paramount, and in times of unforeseen incidents or emergencies, the role of a smart assistant can be particularly reassuring. For instance, in the event of a natural disaster or sudden political unrest, receiving alerts and guidance on safe zones, evacuation routes, or emergency contacts becomes indispensable. The ability to adjust plans swiftly and with confidence can be profoundly comforting.

However, as with any technology, there's an ongoing dialogue about privacy and data security. For effective real-time updates, a tapestry of personal data—location, preferences, and even past travel plans—must be woven together. It's a subtle dance between convenience and consent. Users must remain educated about their privacy settings, ensuring that the data shared is used in ways that respect personal boundaries and uphold data protection standards.

Smart assistants not only help in managing real-time changes but also facilitate proactive planning. They augment our capability to foresee impacts that might not be immediately visible, like seasonal trends or even local events that might affect travel flow and accommodations. By acting as an ongoing travel advisor, a smart assistant transforms what used to be extensive manual research into a streamlined, hands-free experience.

The impact of AI-driven real-time updates on travel is nothing short of revolutionary. By providing travelers with the gift of time, the peace of mind, and the freedom to explore, smart assistants reshape the art of travel planning into something that feels both liberating and luxurious. This capability invites wanderers to challenge traditional travel constraints and step into a world where information flow and adaptability are at their fingertips.

As we continue to embrace and innovate these technologies, the focus will surely shift toward even greater integration with more global services, improved user interfaces, and heightened intelligence capabilities. Future developments might include even more sophisticated predictions and adjustments, driving a continuous evolution of travel where spontaneity and planning harmonize in real-time. With this evolving landscape, travel planning stands at the cusp of even more dramatic transformation.

Chapter 10:
Personalized Entertainment
Experiences

In the ever-evolving landscape of entertainment, AI-powered smart assistants are transforming how we consume media by tailoring experiences to fit our unique tastes and preferences. Imagine coming home after a long day to a playlist that hits just the right mood, or a lineup of must-see movies curated precisely for your weekend marathon. Smart assistants leverage sophisticated algorithms and data insights to deliver personalized recommendations, making our engagement with music and streaming services more intuitive and enjoyable. This intelligent curation not only saves us the time and effort of searching through overwhelming content choices but also introduces us to new genres and creators we might never have discovered otherwise. As AI continues to refine its understanding of our entertainment habits, it promises to enhance our leisure time, ensuring that each moment spent is entertaining and fulfilling in ways we never thought possible.

Streaming Content Suggestions

In today's connected world, where content options might feel endless, getting lost in a sea of television series, movies, and documentaries is all too easy. That's where AI-powered smart assistants step in, offering personalized streaming content suggestions that not only save time but heighten your viewing pleasure. By analyzing your preferences and viewing history, these intelligent assistants can suggest content that aligns closely with your tastes, revealing new titles you might not have considered before.

Imagine settling down on a Friday evening. Your smart assistant knows what genres you adore based on your past queries and watchlists. It doesn't just propose the most popular series or the newest releases; it intelligently curates options that consider what you've enjoyed in the past, weaving in new suggestions that introduce a slight breath of fresh air. The suggestion might be a quirky indie film reminiscent of a director you like or an underrated series from an actor whose performances you admire.

These intelligent recommendations employ advanced algorithms that look into different aspects of your preferences. Beyond just genres and actors, AI assesses subtler elements like pacing, themes, and the overall tone of what you watch. The algorithm takes a multi-dimensional approach, offering a deep understanding of a viewer's unique taste. This kind of personalization feels almost like having a friend who knows your likes and dislikes, ensuring that suggestions are not just accurate but also resonate with your current mood and context.

Furthermore, smart assistants consider various external factors that might affect what you're in the mood to watch. Seasonal changes, special occasions, or even global events can subtly influence your entertainment desires. For instance, around Halloween, you might find suggestions veering towards thrillers or horror comedies. These intelligent nudges suggest that smart assistants recognize that viewer preferences aren't static but rather fluid, responding to the changing world.

Streamlining the selection process means less time scrolling and more time enjoying content. The smart assistant's seamless approach to entertainment reflects a broader trend: technology working to simplify and enhance experience, making interactions more human and less mechanical. And this isn't just limited to single users. Families and cohabitants can benefit from profile-linked suggestions, ensuring that everyone gets to discover relevant content without constant switching and searching.

But how do these systems understand group preferences? Simple yet sophisticated. By assessing overlapping interests and monitoring the frequency of co-viewing, smart assistants provide well-curated lists that cater to collective tastes, embracing a more inclusive approach. This way, family movie nights become more enjoyable, as everyone feels included in the selection.

Privacy remains a top concern when it comes to personalized content. Smart assistants use anonymized data, learning behavioral patterns without compromising user identity. This commitment to privacy allows users to enjoy personalized recommendations while feeling secure. Transparency in how data gets stored and processed remains a pillar of trust in these technological aids.

Looking ahead, streaming content suggestions could evolve beyond traditional media, delving into more interactive formats. Picture exploring immersive virtual environments, curated by your preferences, where you're not just a passive viewer but an active participant. Imagine a world where AI yarns narratives specific to you, transforming not just what you watch, but how you experience stories themselves.

Incorporating social elements, AI can suggest content not just for solitary viewing, but also for social occasions. It might recommend a trending series that everyone at a gathering hasn't seen yet, facilitating an instant shared experience. By leveraging your social network data, smart assistants could connect you with others who share similar content passions, fostering new discussions and connections.

As the integration of AI in entertainment continues to expand, the line between technology and empathy blurs, making experiences feel uniquely personal. Achieving such a balance requires ongoing innovation and ethical foresight. The challenge lies in consistently enhancing algorithms to better mimic human intuition. The more natural these interactions become, the more significant and impactful the role of AI in managing entertainment will be.

Ultimately, the goal is not just about delivering recommendations but enhancing how content enriches our lives. By thoughtfully guiding us through the digital maze of options, smart assistants ensure every moment spent in front of a screen is meaningful and engaging. Partnering with technology, we can unlock a world where entertainment becomes an art of understanding and inspiration, rather than just choice and chance.

Music and Podcast Recommendations

As our lives become increasingly interwoven with digital experiences, one area where smart assistants shine is in tailoring music and podcast recommendations to fit our individual tastes and moods. Picture this: after a long day, you arrive home and casually say, "Hey, play something relaxing." Without skipping a beat, your smart assistant curates a playlist specifically crafted from your listening history and the emotions you typically associate with relaxation. This isn't just convenience; it's the next evolution of personalized entertainment.

Leveraging advanced algorithms and machine learning, smart assistants analyze vast data sets from streaming platforms. They don't just consider what you've listened to recently but take a holistic view of your listening patterns over time. Are you someone who listens to rock on Monday mornings to get pumped up for the week? Or do you prefer jazz during rainy afternoons? These nuances allow your AI to suggest music that you might not have discovered otherwise, expanding your musical horizons while perfectly fitting your context.

Integration with major music streaming services elevates this experience. By linking your assistant to platforms like Spotify, Apple Music, or Amazon Music, it gains the ability to access your playlists, favorite artists, and even your skipped tracks. This integration means the recommendations are not only based on algorithmic suggestions but are also informed by your explicit feedback—likes, replays, and skips. Your assistant becomes more like a personal DJ who's learning your preferences day by day, track by track.

Podcasts, with their diverse range of topics and formats, offer a unique landscape for smart assistants to navigate. Whether you're commuting or simply winding down for the evening, asking for a podcast recommendation could return a wide array of choices, from the latest in tech news to a deep dive into historical mysteries. The beauty lies in the curated balance these assistants achieve—offering both popular podcast episodes and niche content that specifically appeals to your interests.

Imagine diving into a new podcast that's a perfect fit because it was recommended based on your previous interests in particular topics or genres. The AI's sophistication even allows it to adapt recommendations according to your mood, identified either through explicit commands like "I want something funny" or inferred through your habitual usage patterns. It's like having a friend who knows exactly what you want to listen to, often before you realize it yourself.

The underlying technology isn't limited to mere suggestions. Smart assistants can now engage with your listening habits on a deeper level. For instance, they might integrate sentiment analysis into the mix, a capability that allows them to gauge the emotional tone of your preferred content. Thus, when you express wanting to feel motivated, the AI could select energetic, uplifting music or inspiring podcast episodes that align with these subtle cues.

The future of music and podcast recommendations is rich with potential, driven by the continued refinement of AI's understanding of human behavior. This development addresses the challenge of curating personalized content in an era where the volume of available media is both a blessing and a curse. With smart assistants, the overwhelming nature of choice is mitigated by targeted recommendations, freeing you to immerse more fully in the sensory experience rather than the cognitive load of decision-making.

Privacy concerns naturally arise when discussing the AI's capacity to sift through personal data to enhance recommendations. It's crucial to emphasize that this personalization does not come at the cost of security. Modern smart assistants are equipped with robust measures to protect user data while they continually learn and adapt. Opting into these services should always come with clear options for managing data sharing and usage permissions, ensuring that users maintain control over their virtual footprints.

Ultimately, the music and podcasts recommended by smart assistants exemplify the blend of human taste with technological prowess. As you listen to a new playlist or tune into a podcast series that's just right for your drive home, you might pause to appreciate the invisible, tailored artistry of AI at work, shaping and enhancing the soundtrack of your life. In a world brimming with auditory options, your assistant stands as both guide and gatekeeper, ushering in an era of highly individualized aural experiences.

In embracing this technology, individuals explore an interconnected world where entertainment constantly evolves to match personal preferences. The ever-improving capability of smart assistants to sync with your emotional and intellectual needs presents an inspiring glimpse into the fusion of artificial intelligence and human enjoyment. It is not merely about playing music or choosing podcasts—it's an invitation to discover and rediscover the nuanced layers of content that resonate with the chords of your life.

Chapter 11:
Overcoming Privacy Concerns

As smart assistants become more integrated into our daily lives, the issue of privacy can't be ignored. Understanding data security is crucial; these devices collect massive amounts of personal information, making it vital to ensure your data is protected. By adhering to best practices for secure usage, such as employing strong authentication methods and regularly updating software, you can significantly reduce vulnerabilities. It's about finding a balance between convenience and safeguarding personal information, making informed choices that empower users to embrace technology without sacrificing privacy. This approach not only enhances user trust but also paves the way for a smarter, more secure digital future.

Understanding Data Security

In the digital age, data has become one of the most valuable commodities, and it's only natural to be concerned about its security, especially when using AI-powered smart assistants. These assistants seamlessly integrate into our daily lives, but they do so by accessing a treasure trove of personal data. This data is key to providing personalized, efficient experiences that cater to individual needs and preferences. However, as the saying goes, with great power comes great responsibility. Understanding how data security fits into the broader context of smart assistants and privacy concerns is vital for any tech-savvy individual.

Smart assistants collect a significant amount of information, from voice recordings and search history to location data and personal preferences. This is necessary for them to function effectively. They need to understand how we speak, predict what we want, and learn our routines to offer relevant suggestions. Imagine having a digital assistant that helps manage your calendar, sends reminders about important dates, and even optimizes your smart home devices all on cue. However, the processing and storage of such data raise important questions about access, misuse, and protection.

There's no denying the fact that many users feel uneasy about how their data might be used or disclosed. This concern is not unfounded, given the instances of security breaches reported over the years. Hackers continue to evolve their tactics, and unfortunately, no system is impervious. Therefore, the challenge is to stay one step ahead in securing data while embracing the convenience of smart assistants. Developers and tech companies must constantly adapt and innovate their security measures to protect user data against malicious threats.

Encryption plays a pivotal role in securing user data. It is a method that scrambles data, rendering it unreadable to unauthorized parties. Think of it as a locked box that only a person with the right key can open. When your smart assistant stores or transfers sensitive data, encryption ensures that even if data is intercepted, it remains unintelligible. Leading technology companies often employ complex encryption techniques for data both in transit and at rest to maintain a high level of security.

Moreover, data anonymization and pseudonymization are critical processes in data security. These techniques involve modifying data in ways that prevent the identification of individuals, thus deterring unwanted privacy intrusions. By separating personal identities from data analysis, these methods help maintain user privacy while still allowing companies to improve services based on general data trends.

The concept of data minimization also plays a crucial role in understanding data security. This approach encourages collecting only the data truly necessary for the smart assistant to function. In practice, this means that developers should not seek to gather excess information under the guise of "future use" since it creates unnecessary risk. The practice of data minimization respects user privacy and reduces potential vulnerability by keeping only essential data on hand.

User control is an empowering factor in enhancing data security. Many tech companies are recognizing the importance of giving users access to, and control over, their own data. This includes allowing them to view what information is stored, how it's used, and the ability to delete or adjust what's collected. User dashboards and privacy settings are becoming increasingly sophisticated, giving users a stronger grasp on managing their digital footprint.

Transparency is another cornerstone of data security. When companies openly communicate about how they're using data, they build trust with their users. This means clear, understandable privacy policies and regular updates about any changes in data processing practices. Users are more likely to engage with smart assistants confidently when they feel informed and empowered regarding their data security.

Collaborations between tech companies and third-party security experts can help enhance the security strategies of smart assistants. External assessments and ethical hacking initiatives can unearth vulnerabilities that might not be apparent internally. By embracing such collaborations, companies can bolster their defense mechanisms and develop more robust security frameworks over time.

Legislation also plays a significant role in shaping data security measures. As governments worldwide recognize the need to protect user data, new laws and regulations are emerging that define how companies should collect, store, and use data. In anticipation, many companies adopt proactive measures that exceed these legal requirements to safeguard credentials and maintain competitive trustworthiness in the marketplace.

Finally, understanding data security links directly back to user education. While companies and developers strive to implement the best security practices, users must also be vigilant. Regularly updating software, using strong, unique passwords, being cautious about granting permissions, and staying informed about the latest security trends are simple yet effective measures users can take to protect their personal information.

In conclusion, while smart assistants offer unparalleled convenience and smart solutions in our daily lives, maintaining data privacy is crucial. Understanding the keys to data security, such as encryption, user control, transparency, and legislation, enhances our capacity to use these devices responsibly. As technology evolves, so must our commitment to protecting the data that enriches our lives, ensuring a harmonious balance between embracing innovation and securing our digital domains.

Best Practices for Secure Usage

As our lives become increasingly intertwined with smart assistants, ensuring secure usage has never been more vital. Navigating the digital landscape demands vigilance, especially as these devices become storehouses of personal information, guiding almost every aspect of our daily routines. It's imperative to adopt a proactive mindset towards safeguarding data, ensuring both privacy and security are not an afterthought but a primary consideration in our interconnected world.

One of the simplest yet most effective practices is to consistently update the software of your smart assistants. Developers continually patch vulnerabilities and enhance security features through updates, making it crucial to install these as soon as they're available. This simple action often defends against exploits that could otherwise compromise your assistant's integrity. Moreover, regularly updating associated applications on your smartphone or tablet further fortifies this security perimeter.

Utilize robust authentication methods to bolster security. Strong passwords, consisting of a complex mix of alphanumeric and special characters, are foundational to digital security. Encourage multifactor authentication if available, adding an extra layer of protection by requiring something you own in addition to something you know. Biometric solutions, like fingerprint or facial recognition, further enhance secure access to your smart assistant, merging convenience and safety seamlessly.

Consider the physical aspect of security as well. Positioning your smart assistant in a secure location prevents unauthorized individuals from interacting with it directly. Although voice recognition features add a level of security by only responding to familiar voices, these systems aren't infallible. Thus, physical access control becomes an integral part of secure usage practices.

Turning off features you don't use often can be a simple yet powerful security measure. If you're not utilizing certain integrations or capabilities, consider disabling them to minimize potential entry points for malicious actors. Take time to go through the settings and features of your device, understanding each one's purpose and security implications. This not only enhances security but also optimizes your assistant for personal needs.

Another prudent practice involves scrutinizing third-party applications and services connected to your smart assistant. These integrations often serve as extensions to your device's functionality but can also introduce vulnerabilities. Always ensure these applications are from reputable developers and maintain up-to-date security credentials. Regularly review permissions granted to these apps, adjusting or revoking as necessary when certain features or services are not in use.

Additionally, being conscious of voice recordings and data handling policies is essential. Smart assistants rely on voice input, which means they are listening continuously for wake words. Ensuring that your device only retains data necessary for functionality, and understanding how this data is stored or utilized by service providers, is vital. Policies related to data anonymization and encryption should be evaluated before opting into any service.

Education plays a significant role in secure usage. The more users know about both the technology and associated security practices, the more effectively they can protect themselves. This includes understanding potential phishing attacks, which might attempt to capture personal credentials by masquerading as legitimate communications. Ongoing education on recognizing these threats and responding appropriately underpins a strong defensive strategy.

Moreover, auditing your device's usage periodically can offer insights into any unusual activities that might indicate a security breach. Monitor access logs and understand the usual patterns of your device to identify anomalies quickly. Likewise, many smart assistants provide activity summaries that can be reviewed for irregularities or unauthorized access, enabling immediate corrective action.

In the spirit of sustainable practices, ensuring your equipment and network infrastructure are up-to-date and in good operational condition enhances security. Obsolete firmware and outdated hardware can present vulnerabilities that modern devices are better equipped to handle. Additionally, employing secured and encrypted home networks acts as a first line of defense, barring unauthorized access from the onset.

Conclusively, securing the usage of smart assistants in our homes or workplaces demands a blend of vigilance, education, and technology management. As these devices evolve, so do the strategies and tools for safeguarding them. By embedding best practices into everyday use, we can not only enjoy the full scope of their capabilities but do so with peace of mind that our privacy and security are actively protected.

Chapter 12:
Troubleshooting Common Issues

As we dive into the nuances of AI-powered smart assistants, we can't ignore the occasional hiccups users might encounter. From resolving connectivity problems that hinder seamless communication to maintaining device performance in a world pushing the limits of digital efficiency, troubleshooting is crucial. Understanding the nature of these common issues equips us to handle them proactively, minimizing disruption by employing quick fixes and preemptive measures. The evolving landscape of AI technology invites a proactive approach—where adaptability and resourcefulness are key. By learning how to navigate these challenges, users can ensure their smart assistants function optimally, enhancing their daily synergy between human ingenuity and artificial intelligence. With these skills, users can enjoy an uninterrupted digital experience that aligns with their productivity and personal goals.

Resolving Connectivity Problems

One of the most common challenges you might encounter with AI-powered smart assistants is connectivity issues. It's easy to become frustrated when your device is unresponsive or fails to execute commands due to poor network connection. However, understanding and resolving these hiccups can ensure that your interaction with these intelligent assistants remains seamless and efficient.

At the core, connectivity problems between smart assistants and the internet—or even among connected devices—can be attributed to several factors. From outdated software to physical obstructions, the list is extensive. The first step in troubleshooting is always to identify the specific culprit, which often requires a process of elimination. For instance, a simple restart of your router or device might solve many issues. If that doesn't work, you may delve deeper into more technical solutions.

One foundational step in tackling connectivity issues is ensuring your smart assistant's software is up to date. Often, updates not only enhance the functionality of the assistant but also improve its ability to maintain stable connections. It's recommended to enable automatic updates whenever possible, allowing you to benefit from improvements and patches without having to think about it actively.

Wireless interference is another frequent cause of connectivity issues, especially in environments rich with competing signals. You might find that the more devices that are connected to your network, the more strained that network becomes. Consider placing your router in a central location, away from walls or any barriers, for optimal signal dispersion. In some cases, changing your router's frequency band from 2.4GHz to 5GHz can reduce interference.

In environments where signal strength is less than optimal, the deployment of a mesh network could offer a significant enhancement. These systems use multiple nodes to ensure a strong, consistent Wi-Fi signal across your entire space. Given that many smart assistants need to operate with several devices at once, a robust network ensures smooth communication, regardless of where devices are situated.

When troubleshooting, don't underestimate the power of basic solutions. Resetting your router can sometimes rectify complex issues; it's a simple yet effective method that clears up temporary network glitches. Furthermore, checking for loose cables or power disruptions can save you considerable time and frustration.

Device-specific issues also contribute to connectivity problems. If your smart assistant is struggling with maintaining a connection, verify that it is within range of your Wi-Fi network. Additionally, ensure that other devices on the network aren't consuming disproportionate bandwidth, which might leave your smart assistant starved for connectivity.

In some scenarios, the cause of the problem might lie in DNS (Domain Name System) settings, which translate domain names into IP addresses. Changing the DNS server settings in your router to a more reliable or faster option can sometimes alleviate connectivity lags. Public DNS services such as those offered by Google or Cloudflare are popular alternatives.

If you've explored physical and software-based solutions to no avail, it's time to consider potential issues at the service provider level. Occasionally, issues with the internet service provider could affect connectivity, where scheduled maintenance or outages impact your entire network's ability to remain connected. It's worth reaching out to your provider to rule out such possibilities.

For those who find themselves continually battling connectivity issues despite trying everything above, investing in network diagnostic tools or apps could provide deeper insights into your network's health. These tools can monitor and report network usage, enabling you to pinpoint troublesome areas more precisely.

As we increasingly rely on technology to get through our daily routines, connectivity issues remind us of the importance of a resilient infrastructure. But with the right approach and troubleshooting know-how, these problems are often surmountable. Ensuring your smart assistants remain connected is not merely about convenience—

it's a gateway to unlocking their full potential and embedding AI into the fabric of everyday life seamlessly.

So, as you navigate the intelligent world of AI-driven technology, keep these troubleshooting tips at your fingertips. A well-connected smart assistant not only amplifies productivity and convenience but also transforms how we experience digital integration in our every moment.

Maintaining Device Performance

In our rapidly advancing digital age, where AI-powered smart assistants are becoming indispensable to our daily lives, maintaining the performance of these devices is crucial. As we layer more responsibilities onto our digital helpers, from managing schedules to controlling smart home systems, ensuring their optimal operation becomes vital. An efficient smart assistant doesn't just simplify tasks; it seamlessly integrates into our lifestyle, enhancing convenience and efficiency. However, just like any other technology, these devices require care and attention to maintain their peak performance.

First, let's delve into some practical aspects. Smart assistants often rely heavily on internet connectivity to function optimally. Any slowdown in performance might be attributed to connectivity issues. Regularly checking your device's internet speed and stability can prevent unnecessary disruptions. Sometimes, all it takes is a simple reboot of your router or the device itself to get things back on track. Moreover, keeping your device's software up-to-date ensures that it runs on the latest security patches and feature enhancements provided by the manufacturer.

Proactively managing the storage capacity of your device is another important step. Over time, smart assistants accumulate data from various interactions, consuming storage space and potentially affecting device performance. Regularly purging unnecessary data—such as outdated queries, redundant apps, or cached files—can free up valuable system resources. Utilize cloud storage options where available to offload some of the data and maintain local storage space for more crucial operations.

Device performance can also be influenced by the physical environment in which the smart assistant operates. Placing the device in areas with poor ventilation can lead to overheating, which may hamper performance over time. Ensure that the smart assistant resides in a well-ventilated space, away from heat sources or direct sunlight. This simple but often overlooked consideration can extend the lifespan of your device significantly.

The potential of AI-powered assistants isn't solely dependent on their hardware. Their ability to understand, predict, and respond to user commands is continually enhanced by machine learning algorithms. To keep this feature performant, it is essential to regularly review and update user information and preferences. The more accurately your device can predict your needs and preferences, the more seamless the interaction becomes. This dynamic adaptability is one of the defining features of AI and requires the user to provide updated and comprehensive data into their profiles.

Many users underestimate the importance of monitoring the energy consumption of their devices. Leaving smart assistants constantly active may place unnecessary burdens on your home energy systems and inadvertently impact device lifespan. Consider setting specific operating schedules or utilizing features like power-saving modes to balance performance and energy efficiency. Smart assistants, though designed for continuous operation, can benefit from offline breaks during periods of inactivity.

Another facet of maintaining device performance involves ensuring compatibility between your smart assistant and connected devices. Smart assistants often act as the central hub for various smart home gadgets, from lights to thermostats to entertainment systems. Each of these devices receives routine updates that might affect their integration. Regularly checking firmware updates and compatibility information across all connected devices minimizes any potential disruptions in the overall system's function.

In some instances, poor performance can stem from how users engage with the technology. Helping your smart assistant "learn" involves actively refining interaction patterns. Training it by correcting misunderstood commands or experimenting with different phrasings makes it more effective over time. This kind of user feedback is critical, reinforcing positive response patterns and streamlining future interactions. It's an iterative process that strengthens the user-device relationship.

Finally, engaging with online communities and forums dedicated to smart assistants can provide invaluable insights into performance optimization. Other users often share tips, software tweaks, and solutions that might not be readily available in official documentation. These communities can serve as a robust support network, providing fresh ideas and helping troubleshoot any nuances you encounter with your device.

Maintaining device performance for smart assistants is both a commitment and an opportunity. It's about preserving the operational excellence of the devices and enhancing the user experience through personal investment of time and learning. By actively managing various technical and operational factors, users can ensure not only longevity but also the best utility from their smart assistants. The benefits are multifold: a device that operates efficiently saves time, energy, and resources, allowing users to focus more on the human experiences they wish to enrich with the help of AI.

Chapter 13:
Developing AI Skills

Diving into the realm of AI skills, the focus shifts to those keen on harnessing the power of artificial intelligence for personal growth and enhanced productivity. Learning to code with AI isn't just for tech enthusiasts; it's an invitation to engage with technology on a deeper, more intuitive level. Creating custom commands enables users to tailor their smart assistants, making daily interactions more personalized and efficient. This chapter explores how developing these skills can transform an AI assistant into a uniquely capable partner, augmenting not just routine tasks but sparking innovative ways to approach them. By embracing the potential of AI, individuals can elevate their digital fluency, turning challenges into opportunities and crafting smarter, more connected lives in a rapidly evolving digital landscape.

Learning to Code With AI

Embarking on the journey of learning to code can seem daunting, especially for those who haven't had much exposure to programming or technology beyond the everyday use of smart devices. However, with the advent of artificial intelligence, coding has become more accessible than ever before. AI is not just simplifying the process; it's actively reshaping how we approach the learning of this critical skill. In an age where AI is becoming integral to almost every industry, learning to code with the assistance of AI can significantly enhance a tech-savvy individual's proficiency and understanding.

AI-powered tools act like personalized mentors, adapting to your learning pace and style. These tools can offer step-by-step tutorials, instantly pointing out errors in your code and suggesting improvements. For instance, platforms utilizing AI can analyze your code in real-time, ensuring that you learn not just how to fix errors, but also understand the logic behind the corrections. This immediate feedback loop is pivotal in cementing foundational concepts and maintaining momentum in your learning journey.

The advantages of learning to code with AI don't stop at error correction. AI can help map out custom learning paths based on your goals and current skill levels. Whether you're diving into Python for data science or exploring JavaScript for web development, smart assistants can curate resources, exercises, and projects that align with your desired outcomes. This individualized attention transforms coding from a generic, one-size-fits-all endeavor into a tailored educational experience.

Beyond personalized learning paths, AI also drives motivation through gamified learning experiences. Platforms leveraging AI can create challenges and simulations that are both engaging and educational, turning the often dry process of coding into a captivating adventure. This combination of play and work can exponentially boost both engagement and retention of information, making the learning process not only more effective but enjoyable.

Moreover, AI can facilitate collaboration over coding projects, an essential skill for any aspiring developer. Smart assistants can help pair learners with similar interests or complementary skills, forming virtual coding communities. By participating in group projects, learners can experience real-world applications of teamwork and version control, gaining insights that are difficult to acquire through solo study. This collaborative learning reflects the team-oriented nature of most real-world programming jobs.

As you grow more comfortable in your coding endeavors, AI-powered platforms can introduce you to more complex concepts incrementally. They can guide you through advanced topics like machine learning algorithms or neural networks without overwhelming you. The beauty of this approach lies in its flexibility; you can adjust the learning curve to match your comfort and curiosity levels, ensuring that every step forward builds steadily upon the last.

In this rapidly evolving landscape, staying current is crucial. AI tools are continually updated with the latest programming techniques and industry standards, ensuring that your learning material stays relevant. Instead of wading through outdated textbooks or static online courses, learners benefit from a dynamic, constantly evolving educational ecosystem that AI provides.

For those looking to dive deeper into specific niches, AI can also assist in coding research projects. By scouring databases, academic journals, and code repositories, smart assistants can synthesize vast amounts of data quickly, helping you formulate hypotheses and streamline your investigation processes. This capability empowers learners to not only absorb information but also to contribute to the development of innovative solutions practically.

In conclusion, learning to code with AI is about more than simply acquiring a technical skill; it's about harnessing a transformative educational approach. AI brings a level of adaptability and accessibility to coding education that was previously unimaginable. By leveraging AI, you can accelerate your learning while enjoying a rich, supportive, and personalized coding journey. As smart technology continues to integrate into our lives, acquiring these AI-assisted coding skills will only become more critical in both personal and professional spheres.

Creating Custom Commands

At the heart of truly harnessing the power of smart assistants lies the ability to tailor them to your specific needs. This personalization is often achieved through creating custom commands, a feature that elevates your interaction with AI from mere user to an empowered orchestrator of your digital environment. Imagine having a digital assistant that not only responds to generic commands but also understands and executes tasks devised uniquely by you. The potential for increased productivity, convenience, and efficiency is immense.

Creating custom commands can transform a smart assistant into a digital extension of yourself, capable of handling tasks with a level of personalization and specificity that aligns perfectly with your lifestyle. Whether it's setting up a sequence that dims the lights and plays a chosen playlist when you say "Evening Relaxation," or configuring a command that sends a morning briefing to your email, the possibilities are vast. Such automation bridges the gap between technology and daily tasks, offering a seamless transition from intention to execution.

Most smart assistant platforms come equipped with user-friendly interfaces for developing these commands, minimizing the technical know-how required to start. For instance, Apple's Siri Shortcuts, Google's Assistant Routines, and Amazon's Alexa Skills allow users to create simple or complex sequences without delving deep into coding intricacies. These tools are designed to be intuitive, often enabling users to drag and drop actions into a sequence that the assistant will execute. With some platforms, users can also add conditions and triggers that enhance the flexibility and robustness of their custom commands.

However, for those eager to deepen their customization and technical skills, understanding the basic principles of coding can be the key to unlocking even greater potential. By learning a programming language that's compatible with your AI platform, you can craft commands that are as detailed and nuanced as you desire. This doesn't necessarily require becoming a full-fledged programmer; even a rudimentary grasp of coding principles can open doors to advanced customization. Online resources, courses, and community forums are bountiful and ready to assist you along this learning journey.

Creating custom commands involves seeing what parts of your routine can be simplified or automated. Begin by identifying repetitive tasks in your day-to-day activities. Do you regularly adjust your thermostat at certain times, switch between different apps, or need reminders for remote meetings? Make a list and consider how these processes could be woven into a comprehensive command or sequence. Then, use the built-in tools or scripts to layout these tasks logically, test their functionality, and optimize them to ensure reliability.

While setting up custom commands, it's important to think about future scalability. Your daily tasks today might evolve into something completely different a year from now. Creating versatile, robust commands that can be easily modified is an ideal approach. For example, instead of a command that only reminds you about daily meetings, create one that pulls in information from different calendars, adapts to changing time zones, or notifies you about upcoming events through various mediums.

This personalization, though empowering, demands a balanced approach. Over-customization can lead to a maze of complex commands that might become difficult to maintain. Hence, it's crucial to evaluate which commands significantly enhance your quality of life and discard those that merely add to the clutter. It's about finding the sweet spot where technology enhances everyday functions without unnecessary complexity.

In addition to daily task automation, custom commands can serve more creative or entertainment purposes. You might devise sequences that brighten your mood with a joke or interesting fact during certain times of the day, or even commands that activate a whole-home cinematic experience with one phrase. The idea is to leverage technology not just for practicality but also for enjoyment and enrichment.

The impact of custom commands is not confined to individual use; they can vastly improve collaborative environments as well. Businesses can set up commands that streamline workflow, manage project tasks, or facilitate communication across teams. In these settings, custom commands can ensure timely updates, information dissemination, and automate routine business processes, which collectively enhance efficiency and productivity.

Moreover, as smart assistant ecosystems evolve, the interoperability between different devices and services continues to expand. This connectivity further amplifies the power of custom commands. Soon, we may see a unified command that triggers a series of actions across a multitude of smart devices — syncing data for a comprehensive health report or aligning a household's energy consumption with peak savings windows, all with simple phrases.

In conclusion, crafting custom commands not only optimizes time and energy but also fosters an enhanced relationship with technology where it becomes more of a personal assistant than a mere gadget. It represents a shift in how we interact with our digital assistants, transforming them from reactive tools to proactive partners in our lives. Embracing this capability, we unlock new frontiers in personal and professional domains, forging a path towards a more sophisticated and intuitive interaction with technology in our daily lives.

Chapter 14:
Staying Updated with AI Developments

As the landscape of artificial intelligence evolves at an unprecedented pace, staying updated with AI developments becomes essential for anyone eager to harness the full potential of smart assistants in daily life. The rapid advancements in AI technology offer not only innovative tools but also new possibilities for integrating these systems into our routines. By keeping an eye on industry trends, enthusiasts can anticipate changes and adapt swiftly, embracing fresh updates that can enhance productivity and efficiency. Engaging with online communities devoted to AI and smart assistants provides invaluable insights and fosters connections with like-minded individuals. These collaborative spaces often become hubs of innovation, where ideas are shared and solutions to emerging challenges are collaboratively developed. By actively participating in these discussions, users can cultivate a deeper understanding of both the capabilities and limitations of AI, ensuring they remain at the forefront of this digital revolution. Through a proactive approach to learning and adaptation, one can transform the relationship with AI from a mere tool to a true partner in navigating the complexities of modern life.

Following Industry Trends

In a world where technology evolves at an astounding pace, staying updated with the latest industry trends becomes more than a necessity—it's an art. As artificial intelligence and smart assistants continue to reshape our day-to-day experiences, it's crucial to have our fingers on the pulse of this dynamic landscape. But how does one keep up with these swiftly-moving trends?

Following industry trends begins by recognizing the inevitable shifts and transformations within the tech ecosystem. The AI industry, in particular, is notorious for its rapid innovations, each promising to redefine how we engage with technology. For tech-savvy individuals like yourself seeking to harness the capabilities of AI-powered smart assistants, it's about staying informed, understanding where the industry is heading, and adapting to these changes before they become mainstream.

One immediate source of information is the tech giants leading the charge—companies such as Google, Apple, Amazon, and Microsoft. These organizations often set the benchmark for what smart assistants can achieve by constantly improving natural language processing, machine learning capabilities, and user interface designs. By keeping an eye on their latest releases and developments, you gain insights into upcoming features and improvements that could soon become part of your tech toolkit.

The world of start-ups and innovative emerging companies also offers a glimpse into the future. These players introduce groundbreaking ideas and products that often challenge the status quo. Entrepreneurs and developers frequently push the boundaries of smart assistant technology, providing niche solutions that solve specific user needs or enhance existing functions. Participating in tech conferences, workshops, and webinars can provide valuable exposure to these innovations.

In addition to observing the key players, actively engaging with trend-setting publications and blogs is essential. Platforms such as TechCrunch, Wired, and MIT Technology Review consistently provide updates on AI advancements and smart assistant functionalities. They offer in-depth analyses, expert opinions, and forecasts that can help you gauge the trajectory of technological trends. Subscribing to newsletters and following influential tech enthusiasts on social media can ensure a constant stream of relevant information.

Social media remains a powerful tool for trending insight. Twitter hashtags like #AI and #SmartAssistants can connect you to discussions, opinions, and news that shape the tech narrative. Engaging with these platforms allows you to participate in conversations and learn from other professionals, developers, and enthusiasts who are equally invested in the future of AI.

Industry reports from research firms like Gartner and IDC further enhance your understanding by providing data-driven insights and market predictions. These reports delve into the adoption rates, market growth, and potential challenges facing smart assistant technology. By critically analyzing these reports, you can anticipate shifts in the industry landscape and position yourself ahead of the curve.

Moreover, it's important to directly interact with smart assistants in your daily routine. The most profound insights often come from personal experience. By exploring new functionalities, experimenting with updates, and customizing features, you actively participate in the dialogue between user and technology. This hands-on approach fosters a deeper understanding of emerging trends and the practical application of AI enhancements.

Networking within the tech community is another invaluable strategy. Participating in forums, such as Reddit's AI communities, and attending tech meetups and hackathons can facilitate exchanges of insights, experiences, and predictions. These platforms provide opportunities to connect with researchers, developers, and like-minded individuals who share your passion for smart assistant technology.

Having a discernible awareness of regulatory and ethical debates surrounding AI is crucial. The ongoing discussions about data privacy, algorithm biases, and ethical AI usage intertwine with technological advancements. Keeping abreast of policymaker and academic discourse ensures that you approach AI developments with a balanced and informed perspective.

Remember, staying updated isn't solely about absorbing information; it's about fostering curiosity and critical thinking. By questioning how new developments can integrate into and improve your daily life, you're better prepared to utilize these advancements effectively. Through this dynamic process, you transform potential industry shifts into tangible personal benefits.

In conclusion, following industry trends in AI and smart assistants is an ongoing journey. It demands an active, open-minded approach, an ability to adapt quickly, and a willingness to explore uncharted territories. By diligently applying the strategies mentioned, you'll not only stay informed but also inspire innovation within your own life, preparing you for the exciting evolution of AI technologies.

Engaging with Online Communities

In the fast-evolving field of artificial intelligence, staying updated isn't just about consuming the latest news. It's about becoming part of a larger dialogue, and engaging with online communities dedicated to AI often provides unparalleled insights and diverse perspectives. These communities are treasure troves of collective wisdom, where tech enthusiasts, experts, and beginners come together to explore the many dimensions of AI. Engaging with these groups not only keeps you informed but also invites you to contribute to the ongoing narrative about AI and its future.

Online communities offer a dynamic space to discuss the latest in AI developments. This interaction can take many forms, from interactive discussions, forums, and webinars, to virtual meet-ups. Platforms like Reddit and Stack Exchange have become havens for tech-savvy individuals keen on sharing experiences and troubleshooting queries about smart assistants or AI trends. These platforms allow you to tap into the collective knowledge of global experts, and more importantly, pose questions and exchange ideas in real-time.

Whether you're navigating an unexpected challenge with your AI assistant or looking for ways to improve its integration into your daily life, online communities are rich resources. You may find user-generated content, like tutorials and how-tos, that offer fresh perspectives on using AI tools effectively. Users often share firsthand experiences, which can mitigate the learning curve associated with new technological updates. These shared insights can transform frustrations into learning opportunities, allowing for more efficient use of AI assistants.

The diversity within these communities cannot be overstated. They comprise a spectrum of users, including AI researchers, developers, end-users, and even skeptics, all of whom contribute to multi-faceted discussions. This melting pot of opinions encourages critical thinking and challenges conventional understanding, fostering an environment ripe for innovation. By engaging actively, you become part of a broader movement, one that is critically assessing and shaping how AI technologies evolve and interact with society.

Social media also plays a significant role in monitoring AI trends and innovations. Platforms like Twitter and LinkedIn are frequented by industry leaders and influencers who share insights, research, and forecasts about the world of AI. Following these discussions provides a nuanced understanding of how AI is changing the technological landscape in real-time. This is particularly useful for spotting emerging trends such as improvements in natural language processing, advancements in machine learning, and breakthroughs in computational efficiencies.

Online communities can be inspirational, offering glimpses into pioneering projects and use-cases around the globe. This exposure can spark new ideas and encourage you to think creatively about how AI can enhance daily functions. The stories shared within these communities often illustrate the transformative potential of technology, showing how AI is being harnessed creatively to solve complex problems or improve quality of life. This motivational aspect can be highly beneficial, encouraging ongoing learning and curiosity.

Engagement is not just about passively soaking in information; it's about contributing back to the knowledge pool. Sharing your experiences, challenges, and successes with AI assistants can help others in their journeys. Whether it's a unique integration you've managed or a workaround for a common issue, your input is invaluable to the community. This reciprocal model of learning ensures that as AI evolves, so too does the community's collective capacity to harness its potential.

AI-focused online communities often organize webinars and workshops led by leading experts, providing immersive learning experiences unavailable in traditional settings. Attending these events virtually allows you to gain cutting-edge knowledge and methodologies from the comfort of your own home. The interactive nature of these events ensures that learning is a two-way street, enabling participants to ask questions, clarify doubts, and gain deeper insights into AI technologies and their applications.

In essence, becoming an active participant in AI communities helps cultivate a deeper understanding of the technology's potential and its limitations. This proactive engagement develops your ability to anticipate future trends and adapt to inevitable shifts within the AI landscape. Moreover, it strengthens your readiness to leverage smart assistants and other AI tools more efficiently, thereby aligning with the overarching goal of enhancing personal and professional productivity.

Ultimately, engaging with online communities dedicated to AI serves as a vital complement to more formal educational pursuits. It embodies a spirit of open innovation and shared learning that is quintessential to mastering and utilizing AI as a transformative tool. As you integrate learnings from these interactions, your role evolves from a mere consumer of technology to a proactive participant shaping the future of AI.

Chapter 15:
Future of Smart Assistants

A s we look ahead, the future of smart assistants promises remarkable advancements that will redefine our interaction with technology. These intelligent companions are set to become even more intuitive, leveraging enhanced AI capabilities to anticipate and fulfill our needs with minimal input. Imagine assistants that can seamlessly support dynamic work environments, adapt to shifting workflows, and integrate effortlessly into the digital tapestry of our lives. The workplace of tomorrow, fueled by these innovations, will be more efficient and flexible, allowing individuals to focus on creative and strategic initiatives. As AI continues to evolve, smart assistants will transcend beyond mere convenience tools, becoming catalysts for productivity and champions of efficiency in our interconnected world. This transformative journey will not just increase our capacity to accomplish more but also raise significant questions about the balance between technological dependence and human ingenuity, encouraging ongoing dialogue as we craft a future harmonized between humans and machines.

Advancements in AI Capabilities

The future of smart assistants sits at the intersection of rapid technological advancement and everyday practicality. As AI capabilities evolve, smart assistants are not just becoming more efficient but also increasingly intuitive, with a growing capacity to understand the complexities of human behavior. At the core of this evolution is the ever-improving ability of AI systems to process and analyze vast amounts of data swiftly and accurately. This is paving the way for smarter decision-making and proactive assistance across various aspects of daily life.

Machine learning models have become more sophisticated, with systems now possessing the capability to learn from user behavior over time. This adaptive intelligence allows smart assistants to offer a personalized experience that aligns closely with individual preferences and routines. For example, through pattern recognition and predictive analytics, smart assistants can anticipate needs before they are explicitly expressed, suggesting actions such as reminding a user to leave for an appointment based on current traffic conditions. This shift from reactive to proactive assistance is a leap forward in AI's role in everyday life.

Natural language processing (NLP) has seen remarkable improvements, allowing smart assistants to engage more seamlessly and naturally in conversations. The development of advanced NLP algorithms means that these assistants can now understand context, sentiment, and even subtle nuance in human speech. This translates to more meaningful interactions where the assistant can discern, for instance, when a user is frustrated or in need of additional information. As they become more conversational and capable, smart assistants are transcending their previous limitations to offer a level of engagement that feels human.

Another area showing tremendous progress is the integration of AI with Internet of Things (IoT) devices. As smart home technology expands, the synergy between smart assistants and IoT devices is revolutionizing home automation. Households are experiencing unprecedented levels of control and coordination of everything from lights and thermostats to security systems and kitchen appliances. This connectivity allows smart assistants to create an ecosystem that operates smoothly and efficiently, optimizing resource usage and enhancing convenience.

The use of AI in smart assistants also extends into realms like emotional intelligence and empathy. Research is underway to endow AI systems with the ability to recognize and respond to human emotions. By analyzing voice tone, facial expressions, and physiological signals, future smart assistants could offer support and companionship to individuals experiencing emotional distress. This progression towards empathetic AI presents opportunities to improve mental health support and foster more intimate connections between humans and machines.

In industries such as healthcare and education, the advancements in AI capabilities carry transformative potential. Personal health monitoring systems already rely on AI to track and analyze health metrics, providing actionable insights to users and medical professionals. Similarly, in educational settings, AI can deliver personalized learning experiences by adapting material and feedback to the unique learning pace and style of each student. These intelligent systems represent a monumental shift towards individualized service in these crucial sectors.

Security and data privacy remain at the forefront of AI advancements as well. As these systems become more sophisticated, the need for robust security protocols is critical. Advances in encryption and anonymization techniques aim to safeguard user data, ensuring that the benefits of smart assistants are not enveloped in risks. Developing trust through transparency in data use and control is essential in nurturing a harmonious relationship between users and their digital counterparts.

The role of AI in environmental sustainability is also noteworthy. Smart assistants equipped with advanced AI can significantly contribute to energy management and sustainable practices. By intelligently optimizing energy consumption in homes and enterprises, AI can play an essential role in the wider agenda of combating climate change. These systems streamline processes such as automatically adjusting heating and cooling based on usage patterns, contributing to greener and more sustainable living.

Moreover, we are witnessing a gradual blurring of lines between distinct smart assistant platforms. With future developments, interoperability will become more seamless, allowing various species of devices and virtual assistants to work collaboratively. This interconnectedness will enhance their ability to share data, coordinate tasks, and provide a cohesive user experience across different environments and devices.

Finally, as AI capabilities advance, there is a growing focus on democratizing access to this technology. The development of user-friendly interfaces and accessible platforms aims to empower a broader population to harness the power of smart assistants. This inclusivity fosters an environment where technology can be leveraged to improve quality of life, irrespective of an individual's technical expertise or background.

In summary, advancements in AI capabilities are driving smart assistants towards becoming indispensable partners in daily life. By integrating enhanced understanding, empathy, and connectivity, these systems are poised to usher in an era where technology is seamlessly intertwined with human needs and desires. As these innovations unfold, we stand at the cusp of a transformation that promises to redefine efficiency, convenience, and the way we navigate our interconnected world.

Impact on Future Workplaces

The modern workplace is on the cusp of a transformation that promises to redefine productivity and collaboration as we know it. Smart assistants, powered by advances in artificial intelligence, are not just tools but partners that will reshape how we approach our work. These digital aides are becoming more than just convenience tools; they're evolving into integral components of the workplace ecosystem, offering novel solutions to age-old challenges.

Imagine entering an office where mundane tasks are no longer part of your daily grind. Smart assistants can handle scheduling, send reminders, and even arrange meetings autonomously, allowing employees to focus on work that truly requires their creativity and human touch. This shift will fundamentally alter how time is allocated in offices, potentially leading to a decrease in work-related stress and an increase in job satisfaction. As these assistants continue to learn and adapt, their capacity to predict user needs will only get sharper.

The decentralized nature of the modern workforce, accelerated by the rise of remote work, underscores the importance of seamless communication. Smart assistants will act as mediators across various platforms, ensuring that no message is missed and that collaboration flows smoothly, regardless of geographical barriers. Picture a virtual assistant that integrates all your communication channels and organizes them based on priority, sentiment, and urgency. This function alone can alleviate communication overload, one of the significant causes of burnout in the digital age.

However, the impact of smart assistants on future workplaces isn't one-dimensional. They promise to play a pivotal role in skill development and personalized learning as well. By analyzing user interactions and work habits, these assistants have the potential to offer tailored training modules and growth path suggestions. This evolving professional development tool will ensure that employees remain engaged, continuously learning, and aligned with career aspirations, making work more fulfilling.

The integration of smart assistants within software systems extends beyond mere utility. Businesses can leverage this technology to enhance data management systems, optimizing everything from supply chain logistics to customer relationship management. Imagine software that doesn't just collect and present data but offers insights into patterns and projections, all driven by AI-backed smart assistants. This kind of strategic assistance could lead to unprecedented efficiencies in operations.

Of course, while the enthusiasm for these advancements is understandable, it also comes with a renewed focus on data privacy and ethical considerations. Employers will need to ensure transparent data handling practices, giving employees confidence that while smart assistants observe and learn, they do so with privacy at the forefront. Navigating this balance will be crucial in fostering trust and maintaining ethical integrity within organizations.

Another area where smart assistants promise to make significant waves is in inclusivity. As workplaces become more diverse, these assistants can ensure that different languages and accessibility needs are addressed. With AI's ability to process and translate languages and dialects in real time, communication barriers that once seemed inevitable will be overcome, enriching collaborative efforts across multicultural teams.

Furthermore, in decision-making processes, AI can serve as an impartial advisor. By removing human bias, smart assistants can provide data-driven recommendations, thus enhancing decision-making efficacy. Whether it's hiring new talent, assessing employee performance, or evaluating potential business ventures, AI-driven insights can substantially elevate the objectivity and fairness of these decisions.

As automation continues to reduce the need for conventional roles, the reconfiguration of job descriptions becomes inevitable. Smart assistants will not only aid in alleviating redundant tasks but will, in turn, create new roles focused on managing, programming, and improving these systems. The emergence of AI-centric roles will demand a workforce that is proficient not only in working alongside AI but also in understanding its broader capabilities and implications.

To prepare for these changes, educational systems and companies alike must invest in foundational AI knowledge. This educational push will enable employees to operate confidently in a new workspace landscape and facilitate a smoother transition to AI-enhanced roles. The democratization of AI knowledge, driven by smart assistants capable of teaching and guiding, can create a more equitable playing field, fostering innovation from all corners of the workforce.

Organizational culture will also undergo a metamorphosis. As AI takes over more routine aspects of work, culture will shift towards valuing human-centric skills such as emotional intelligence, creativity, and strategic thinking. Leaders will need to focus not on mere oversight of tasks completed but on inspiring teams to harness their human potential in ways that complement the capabilities of smart assistants.

As we look to the future, the partnership between humans and AI in the workplace stands to bring about unprecedented levels of innovation and efficiency. Like any significant change, the path forward will not be without its challenges. However, embracing an AI-augmented future with open minds and adaptable strategies promises a workplace environment that is not only more productive but is also more fulfilling and dynamic.

The journey with smart assistants doesn't stop here, and we must stay curious and forward-thinking as these intelligent helpers continue to evolve. As we collectively shape the future of work, embracing the symbiotic relationship between human potential and AI capabilities will be crucial in building the workplaces of tomorrow. The transformative era of smart assistants is already at our doorstep, and it's up to us to craft the future where technology truly empowers humanity.

Chapter 16:
Balancing Technology and Human Interaction

In a world increasingly mediated by technology, the pivotal challenge lies in harmonizing the efficiency and convenience brought by smart assistants with the fundamental need for genuine human interaction. While these AI-powered companions are adept at streamlining tasks and enhancing productivity, there is a subtle art in curbing overreliance to preserve the essence of real-world relationships. It's about finding equilibrium where technology supports rather than supplants, ensuring that digital engagements expand our capabilities without diminishing our human connections. As we venture further into this digitally connected era, deliberate steps are required to cultivate environments where personal interactions maintain their richness amidst the backdrop of technological advancement. Whether it's through setting boundaries for digital use or fostering face-to-face communication, achieving this balance is key to leveraging the benefits of AI while nurturing our innate desire for human touch and presence.

Avoiding AI Overdependence

In the vast labyrinth of modern technology, smart assistants have emerged as pivotal tools that streamline the complexities of daily life. As we integrate these marvels of artificial intelligence into our routines, it becomes increasingly important to address the issue of avoid AI overdependence. While these digital aids simplify tasks—from juggling calendars to managing household tasks—over-reliance can lead to a host of pitfalls that subtly erode our autonomy and human interaction.

Picture a morning where a smart assistant handles everything: it turns on your lights, brews your coffee, and summarizes your day's schedule. On the surface, this is convenient. Dig a little deeper, though, and you might find a disconnect creeping into the simple joys of life. Dependence on these devices might strip away moments that encourage self-reliance and diminish our capacity to problem-solve independently.

One of the main concerns with AI overdependence is the potential for skill atrophy. Skills that once required active involvement risk being diminished or even lost when delegated to machines. Consider navigation, for instance. Relying solely on a smart assistant for directions could weaken one's spatial awareness and map-reading skills. Similarly, entrusting language translations entirely to an AI may reduce our motivation to learn new languages—a process that enriches cognitive ability and cultural understanding.

Furthermore, heavy reliance on AI-driven decision-making might curb our critical thinking abilities. In pursuit of efficiency, there's a tendency to offload decisions to algorithms, be it for choosing a restaurant or managing investment portfolios. However, this practice could limit our exposure to diverse viewpoints and experiences, resulting in a narrowed mindset. To maintain a robust decision-making ability, it's crucial to blend AI advice with human intuition, ensuring that technology complements rather than replaces our judgment.

Moreover, there's an inherent risk in becoming too dependent on technology when systems fail or encounter limitations. We've all experienced the frustration when our digital assistants are unresponsive during outages or technical glitches. By centering our lives around them, we become vulnerable to these disruptions, highlighting the need for backup plans. Rediscovering skills like time management, cooking without digital instructions, or balancing a checkbook can buffer us against these interruptions.

Avoiding AI overdependence doesn't mean rejecting technology; instead, it calls for a balanced synergy between human endeavor and artificial intelligence. Embracing tech prudently means setting boundaries—designating certain activities as tech-free zones, like mealtimes or bedtime, which fosters meaningful human interaction. Such practices reinvigorate face-to-face communication, a vital element often overshadowed by our penchant for digital convenience.

Another strategy involves engaging deeply with technology to understand its workings beyond superficial usage. By learning the fundamentals of how AI operates, such as delving into basic programming or exploring AI functionalities, we transform from passive users to empowered ones. This active engagement not only enhances our appreciation for AI capabilities but also equips us with insights into its limitations, prompting us to use it more judiciously.

The dichotomy of technology and humanity isn't a zero-sum game. There's an inspirational middle ground where technology amplifies our humanity rather than diminishes it. Companies and individuals alike are increasingly exploring ways to encourage user interactions that avoid overdependence. Initiatives that promote tech-literacy and workshops on mindful usage of digital assistants are steps towards sustainable technology integration.

At the core of balancing this relationship is a profound understanding of what it means to be human in a digitized world. As we navigate the boundaries of AI influence, reflecting on the purpose and impact of each task we delegate to a machine can inspire a more intentional relationship with technology. By maintaining our educational pursuits, fostering creativity, and valuing interpersonal interactions, we can ensure that technology serves us, not the other way around.

Ultimately, leveraging smart assistants should be about enhancing our capabilities, not diminishing them. The future, rich with the promise of advancement, beckons us to be innovators and thoughtful stewards of the technology we embrace. Finding harmony in this balance will pave the way to a smarter, more connected world that doesn't overshadow the quintessential human spirit.

Promoting Real-World Connections

In our increasingly digital world, the role of smart assistants has never been more pivotal in bridging technology with human interaction. While these intelligent companions harness the power of AI to simplify our daily routines, they also offer invaluable opportunities to strengthen real-world connections. The challenge lies in leveraging these technologies in ways that complement, rather than detract from, the richness of human experiences.

At the heart of this dynamic is the ability of smart assistants to handle mundane tasks. By automating simple routines, they free up time that can be invested in meaningful interactions with family, friends, and colleagues. Imagine coming home from a long day, and instead of grappling with a never-ending list of chores, your smart assistant seamlessly handles grocery orders, sets your favorite playlist, or even adjusts the thermostat to your liking. This don't just deliver convenience; it affords us the luxury to focus on people who matter.

Yet, promoting real-world connections isn't just about reclaiming time. It's about using these digital tools to facilitate and enhance our interactions. Consider the use of smart assistants in organizing events or coordinating schedules—a task that's often fraught with logistical nightmares. With AI, it becomes remarkably easy to find common availability, send reminders, or even book venues. This means less time worrying about logistics and more time enjoying the presence of company.

Moreover, smart assistants have the potential to deepen connections by delivering personalized experiences. They can recommend activities, dining spots, or travel destinations based on your preferences. It's as if having a personal concierge, one who knows precisely what you love and tailors suggestions accordingly. This personalized approach not only helps you create memorable experiences but also enriches interactions with loved ones, as shared interests and activities become easier to discover and pursue together.

Nonetheless, the real magic happens when smart assistants are used as tools to foster communication rather than replace it. In professional settings, AI can enhance collaboration by simplifying the coordination of tasks and projects, thereby allowing individuals to focus on the core of their work—collaborating and innovating. By distributing the workload of mundane tasks, colleagues can engage more deeply in creative problem-solving and strategic discussions.

The integration of smart assistants doesn't stop at personal or professional lives; it extends into community interactions. With features that aggregate local news, alerts, and event information, AI can be instrumental in encouraging community engagement. Whether it's discovering a neighborhood event or participating in local volunteer opportunities, smart assistants can highlight ways to be actively involved with others around us, leading to stronger community bonds.

However, as we navigate this landscape, we must remain vigilant against the potential pitfalls of over-reliance on technology, which can inadvertently lead to social isolation. Smart assistants should act as facilitators of interaction, not barriers to it. Encouraging face-to-face conversations, fostering genuine empathy, and maintaining authentic relationships are skills that must continue to be nurtured in the age of

AI. By setting intentional boundaries on how and when we use technology, we can ensure that our devices serve as bridges rather than walls.

The inspiring part of this journey is recognizing the balance AI can strike between efficiency and empathy. Our task is to ensure that smart assistants enhance our ability to connect by freeing us from the rigmarole that often crowds our attention. This means regularly evaluating personal and professional priorities, making sure they don't fall by the wayside as we embrace new technologies.

Furthermore, fostering real-world connections also involves acknowledging the need for inclusivity. By designing these technologies to accommodate diverse needs and ensuring accessibility for all, we are moving towards a future where everyone has the opportunity to engage in social interactions—regardless of physical or cognitive abilities. This democratizing power of AI has the potential to draw us closer together as a society.

In conclusion, smart assistants hold the promise of reconnecting us with the tangible world by optimizing the intangible. By consciously integrating these digital tools into our lives, we can create a harmonious blend of technology and human interaction, each complimenting the other. As we continue to explore the potential of AI, our goal should always be to use it in ways that enrich our interactions and deepen our connections, creating a world that is not only technologically advanced but emotionally vibrant.

Chapter 17:
Smart Assistants for Accessibility

In the realm of modern technology, smart assistants are dramatically reshaping the landscape of accessibility, making digital experiences more inclusive for individuals with disabilities. By leveraging advanced AI algorithms, these smart assistants go beyond mere voice recognition; they engage in meaningful interactions to provide tailored support across various platforms. One can't help but be inspired by how these tools dismantle barriers, empowering users through real-time transcription, visual assistance, and intuitive navigation. By fostering autonomy and promoting equal access, they're not just gadgets but gateways to participation in an increasingly digital world. The ripple effect of such technological evolution is profound, touching lives by transforming obstacles into opportunities and setting a precedent for future innovations aimed at driving social equity. This powerful integration of AI into accessibility tools is paving the way for a society where inclusivity is the norm, and every individual can fully embrace the digital age with confidence and ease.

Enhancing Digital Accessibility

In an increasingly digital world, accessibility is not just a buzzword; it's a necessity. The rise of smart assistants is promising new horizons for individuals with disabilities, bridging gaps that were once insurmountable. These AI-powered tools are not merely conveniences;

they are critical lifelines that foster independence, inclusivity, and empowerment.

Smart assistants, like Apple's Siri, Amazon's Alexa, Google Assistant, and Microsoft's Cortana, have transformed the landscape of digital accessibility. They offer voice-controlled interfaces that can perform an array of tasks—allowing users to dictate messages, control smart home devices, manage schedules, and even access detailed internet searches without ever needing a screen. For someone with visual impairments, this voice interaction can elevate their everyday experiences by turning complex navigation into a conversation.

Consider the power of these technologies in providing accessible navigation. GPS systems integrated with voice assistants can guide visually impaired individuals through busy streets using voice feedback and auditory cues. This not only aids in their mobility but also enhances their confidence in exploring environments that were previously daunting.

Moreover, speech-to-text technologies embedded within smart assistants are invaluable for those with hearing impairments. By converting spoken language into text, these tools ensure that conversation and information are accessible, whether it's transcribing meetings or interpreting in real-time. Such advancements in real-time recognition and processing are making communication seamless across different mediums and contexts.

The customization capabilities of smart assistants are worth noting too. Users can tailor these devices to meet their specific accessibility needs, whether through personalized command configurations or sensitivity adjustments in voice recognition. This adaptability ensures that individuals with different disabilities can configure their smart assistant in ways that maximize its utility and inclusivity.

Accessibility enhancements extend beyond physical disabilities. Those with cognitive impairments also benefit from these tools. Smart assistants can break down complex information into manageable chunks, remind users of daily tasks, or facilitate structured routines. By providing gentle time prompts and reminders, these devices help manage schedules and reduce cognitive load, allowing users to focus on essential tasks without feeling overwhelmed.

The integration of artificial intelligence in smart assistants has led to innovations like predictive algorithms that anticipate user needs and offer suggestions based on past behaviors. For those who struggle with organizational challenges, whether due to ADHD or other cognitive conditions, this capability offers a significant boost in maintaining focus and structure in daily activities.

Yet, the power of enhancing digital accessibility isn't solely in the hands of tech giants. There is an exciting landscape for developers and innovators to create custom applications that further the reach of accessibility features on these platforms. Open and interoperable APIs provide opportunities to develop specialized apps tailored to niche accessibility needs, fostering a community-driven approach to innovation.

As we explore the future potential of smart assistants in accessibility, the conversation must also include ethical considerations. How privacy is maintained will be crucial since such devices collect and process personalized data to function effectively. Balancing convenience with security ensures that we do not compromise users' personal information while offering these transformative benefits.

Moreover, the evolution of smart assistants for accessibility must be inclusive in itself, considering the diverse range of disabilities. Collaboration with disability advocacy groups in the design and testing phase can ensure that the technology meets practical needs and avoids unintended barriers that might result from assumptions about user capabilities.

Finally, governments and institutions play a pivotal role in championing these technologies. By enforcing and encouraging standards for accessibility, they can ensure that innovations in smart assistants align with inclusive design principles. Such measures can also drive the proliferation of these technologies, making them more affordable and reachable to the masses, further democratizing access to vital digital resources.

In conclusion, smart assistants hold transformative potential in enhancing digital accessibility, providing independence and inclusivity to millions. They serve not only as technological marvels but as tools for societal change, empowering individuals to navigate the digital world with unprecedented ease. As technology continues to advance, the true challenge lies in harmonizing innovation with empathy and responsibility, ensuring that these powerful devices continue to support and uplift all members of society.

Tools for Individuals with Disabilities

Smart assistants have become an integral part of improving accessibility for individuals with disabilities, offering a range of tools that ensure technology is inclusive and empowering. Through voice recognition, facial recognition, and context-aware capabilities, these assistants provide significant support to overcome the barriers people with disabilities often face.

For those with visual impairments, smart assistants offer features that can describe surroundings, read aloud text from various devices, and even decipher currency. The ability to recognize objects or people through device cameras and describe them using natural language processing can be a game changer. These features enable users to gain more independence, as they navigate their daily environments with greater confidence and ease.

Individuals with hearing impairments can benefit greatly as well. With smart assistants, they can receive visual notifications and captions for audio content, ensuring they don't miss any vital information. Devices can transcribe conversations in real-time, allowing users to participate more fully in social and professional interactions. This kind of direct, visual communication supports a seamless exchange of ideas and information, minimizing isolation and enhancing their connectedness to the world.

For those with mobility impairments, smart assistants serve as vital tools that offer hands-free control over various aspects of their life. Voice command functionality allows for easy control over smart home devices, such as adjusting lighting, managing climate control, and controlling entertainment systems without needing physical manipulation. This technology facilitates a more accessible and convenient living space, prioritizing the independence and personal comfort of users.

Cognitive disabilities also receive a boost from smart assistants. These devices can help by breaking down complex tasks into manageable steps, providing reminders, and offering a structured environment for task completion. They can offer cognitive assistance by managing schedules, offering verbal cues, or suggesting routines tailored to individual needs. By creating an environment of consistent support, users experience decreased stress and increased self-reliance.

Emotionally, these tools also provide a supportive presence, offering companionship through conversational interaction. Smart assistants can engage users in meaningful dialogue, offer mental health support through meditative exercises or even remind users to take breaks during stressful times. They transform into a comforting entity, reducing feelings of loneliness and providing gentle stimuli that can improve mental well-being.

The integration of AI into accessibility tools transcends personal devices, extending into public spaces as well. Public transit systems, stores, and public buildings are adopting these technologies, creating a more universally designed environment. With these systems in place, individuals with disabilities can travel more freely, complete transactions more easily, and participate in public life without feeling the limitations previously imposed by inaccessible designs.

However, it's essential to ensure that these advancements remain inclusive. Building AI that understands the nuances of various disabilities requires additional training datasets and sensitivity in design. The stakeholders developing these technologies must include individuals with disabilities to ensure tools are not only functional but also respectful and considerate of their specific needs. Engaging these communities in the design process enriches the potential of smart assistants, ensuring they evolve into truly inclusive technologies.

As we look forward, the continuous improvement of AI technologies involving machine learning algorithms and adaptive functionality promises to offer even more robust solutions for individuals with disabilities. With every iteration, these tools become smarter and more finely tuned to individual nuances and preferences, transforming lives by crossing over existing barriers and promoting a more inclusive and equitable technological landscape.

In conclusion, smart assistants are profoundly changing how individuals with disabilities interact with the world around them. They serve as tools of empowerment, offering solutions that cater to a wide array of needs and ensuring technology is universally accessible. This evolution underscores an inspiring shift towards equality where technological growth is synonymous with inclusivity, creating a more accessible future for all. These advancements form a beacon of hope and progress, demonstrating AI's true potential in enhancing the human experience on an individual level.

Chapter 18:
AI in Shopping and Retail

In a world increasingly driven by digital innovation, AI is transforming the shopping and retail landscape at a breathtaking pace. Imagine walking into your favorite store, whether virtual or physical, and being greeted by a tailored shopping assistant that knows your preferences, anticipates your needs, and guides you seamlessly through your purchasing journey. With smart algorithms at the helm, personalized shopping experiences are now more accessible than ever, analyzing past behaviors and real-time data to recommend products just for you. Meanwhile, the checkout process has evolved from a queuing chore into a frictionless transaction. Self-checkouts powered by AI and mobile apps equipped with sophisticated scanning and payment capabilities are paving the way for a truly seamless retail experience. As technology progresses, the amalgamation of AI with retail doesn't just enhance convenience; it redefines how we perceive consumer engagement, fostering a shopping environment that's intuitive and remarkably efficient.

Personalized Shopping Assistance

In today's retail landscape, personalized shopping assistance is not just a nice-to-have; it's transforming how we interact with products and brands. Artificial intelligence stands at the center of this transformation, providing insights and suggestions tailored to the individual. This integration of AI into shopping has reshaped consumer expectations and experiences. It's no longer sufficient for a brand or retailer to offer a generic shopping experience. Customers crave personalization in everything from product recommendations to promotional offers.

AI-powered shopping assistants serve as digital concierges, learning preferences and making recommendations about what to buy, when to buy, and even how to save money. They analyze past purchase behavior, browsing history, and even social media activity to provide tailored advice. Consider a typical scenario: you're thinking of purchasing new shoes for an upcoming event. An AI assistant can suggest styles based on your past purchases or current fashion trends, notify you about discounts at your favorite retailers, and even handle logistics like ordering and delivery.

The capabilities of AI in personalized shopping aren't confined to suggestions alone. They extend into price comparisons, finding the best deals, and strategically shopping based on individual budget constraints. AI can scan thousands of websites in real-time, ensuring you're getting the best value for your money without the need to open countless tabs or check multiple apps. Such tech capabilities make shopping not only more efficient but also more financially savvy.

This seamless shopping experience is largely driven by sophisticated machine learning algorithms. These algorithms allow AI to rapidly process large volumes of data and swiftly adjust to consumer behavior changes. Through constant learning, AI can identify patterns and anomalies, ensuring that the recommendations remain fresh and relevant. This continuous evolution is central to maintaining customer engagement and enhancing their shopping journey.

Let's delve into the role of virtual shopping assistants. They provide a bridge between e-commerce platforms and consumers, offering guided product tours, answering questions, and facilitating a truly immersive shopping experience. For instance, when considering a high-end product like electronics or home appliances, customers often have numerous questions. Here, an AI assistant can offer insights into specifications, compatibility, reviews, and pricing history, essentially serving the role of a knowledgeable in-store associate.

Moreover, as consumers increasingly rely on mobile devices, AI assistants are there to enhance the mobile shopping experience. They offer features like augmented reality previews, which allow customers to visualize products in their homes before making a purchase. This is particularly beneficial in areas like fashion and furniture, where seeing an item in context can make or break a buying decision.

But it's not just about making the sale. AI shopping assistants are adept at nurturing relational aspects with customers. By managing wish lists, tracking upcoming deliveries, or even scheduling store visits, they work to enhance brand loyalty. A personalized touch, facilitated by AI, can transform a one-time shopper into a repeat customer. This relationship-building is crucial in an era where consumers are bombarded with choices and distractions.

The integration of AI in personalized shopping also raises questions about privacy and data security. While AI assistants offer incredible conveniences, they rely heavily on accessing and analyzing personal data. Ensuring customer data is protected while providing a seamless personalized experience is an ongoing challenge for tech companies. These firms must balance providing valuable services with maintaining consumer trust.

As AI continues to evolve, so will its role in enhancing shopping experiences. Future innovations will likely see even more predictive capabilities, where AI can anticipate needs before they're expressed. Imagine receiving notifications about a sale on a favorite brand before you even knew you needed it, or having an assistant automatically curate your wardrobe based on seasonal changes and personal style evolution.

Moreover, personalization is expanding beyond product recommendations. Customer service is increasingly being augmented by AI, providing personalized support and resolving issues on demand. Virtual assistants equipped with natural language processing can comprehend and process complex customer queries, recommending solutions, or escalating to human agents when needed.

In conclusion, the power of AI in personalized shopping assistance lies in its ability to anticipate, adapt, and augment the shopping experience beyond mere transaction facilitation. It offers meaningful interactions that respect personal preferences and deliver relevant service, evolving as users' technological and lifestyle landscapes continue to shift. As AI's footprint in retail expands, the boundary between exploring and purchasing blurs, creating a deeply intuitive shopping journey.

Seamless Checkout Experience

The evolution of artificial intelligence in the retail sector is rewriting the rules of customer engagement and satisfaction. As we dive into the subtleties of a seamless checkout experience, it's crucial to understand the unprecedented levels of convenience and efficiency AI brings to shopping. Whether walking down a store aisle or browsing through a digital marketplace, AI ensures that once a decision is made, the transaction is swift, secure, and straightforward.

Traditionally, the checkout process has been riddled with bottlenecks and mundane routines. Lengthy queues, multiple verifications, and payment method frustrations have often marred the shopping experience. Enter AI-driven solutions, which are here to alleviate these outdated hurdles. Smart checkout systems are transforming the shopping landscape, making transactions not just faster but also more intuitive. These technologies empower consumers by simultaneously streamlining interactions and enhancing user control over their purchasing processes.

One of the remarkable applications of AI in retail is computer vision technology, which makes cashier-less shopping a reality. By using AI-driven image recognition software, stores can track items selected by customers and accurately identify when they're picked up or returned to the shelf. This technology, often paired with RFID tags or advanced sensor arrays, allows customers simply to walk out with goods, confident that their accounts are correctly charged without the need for scan-and-pay. Not only does this increase efficiency, but it also reduces operational costs and minimizes human error.

Privacy, however, is a key concern when implementing AI technologies. Retailers must walk the fine line between leveraging customer data to enhance experiences and ensuring user privacy is respected. Educating consumers about how their data is used and safeguarded can foster trust and wider acceptance of these new systems. Moreover, giving customers control over their data, such as opting into data usage policies, can serve as a foundation for a long-term trustworthy relationship between retailers and consumers.

AI also plays a significant role in personalized checkout experiences through the use of machine learning algorithms that analyze past purchase behavior. By anticipating customer needs, shopping assistants can suggest relevant add-ons or alternative products as a part of the checkout process. This not only enriches the customer experience but also opens up additional revenue streams for businesses. Imagine a digital shopping cart that instinctively updates with forgotten essentials or offers a discount on frequently bought items, all through AI predictions.

Contactless payment solutions have seen explosive growth, dovetailing perfectly with AI-driven stores. Near Field Communication (NFC) technology, coupled with mobile wallets and

AI-enabled payment apps, allows for swift, touch-free transactions. AI enhances this system by managing and securing these exchanges, identifying unusual patterns, and preventing potential fraud. The beauty of integrating AI with payment systems lies in its ability to make intelligent decisions in milliseconds, ensuring the process remains seamless and secure.

A further extension of AI in seamless checkout is the use of voice-enabled transactions. Voice assistants play an increasing role in the shopping journey, from product search to payment confirmation. By processing natural language queries, these AI systems remove the complexity of navigating multiple layers of a digital storefront. Voice commerce is on the rise, making shopping as simple as telling your device what you need, validating its addition to your cart, and confirming payment—all through natural conversation.

Future advancements in AI may continue to blur the boundaries between physical and digital retail environments. Augmented Reality (AR) and Virtual Reality (VR) holdings are poised to complement these AI tools by offering immersive experiences right at the point of sale. As these technologies mature, customers may virtually try on clothes, sample products, or even walk through a virtual store, with AI ensuring that the checkout, whether actual or virtual, remains an unparalleled, fluid experience.

In summary, the application of AI in creating a seamless checkout experience exemplifies how technology, when applied thoughtfully, can turn tedious tasks into enjoyable interactions. The various AI implementations—from computer vision to voice payments—collectively enhance retail shopping, making it more accessible and personalized for the consumer. As technology advances, the onus will be on businesses to balance innovation with responsibility, ensuring that the benefits of these groundbreaking systems are experienced universally and ethically.

Chapter 19:
Environmental Impact of AI

The intersection of artificial intelligence and environmental sustainability is becoming increasingly relevant as AI technologies continue to embed themselves in our daily lives. While these smart assistants simplify tasks and boost our efficiency, they're also reshaping our ecological footprint. One of the pressing concerns is AI's energy consumption—training and operating complex AI models demand significant computational power, which, in turn, requires considerable energy resources. However, AI-driven technologies also stand at the forefront of helping us manage and reduce energy use. For instance, smart systems optimize energy efficiency in both residential settings and large-scale industrial processes, reducing waste and promoting greener habits. Moreover, AI is aiding sustainable practices by fostering innovations like precision agriculture, optimizing supply chains, and even conserving biodiversity through predictive analytics. While we enjoy the conveniences AI brings, it's crucial to stay mindful of its environmental implications and harness its potential to drive meaningful change toward sustainability.

Energy Consumption Management

Artificial intelligence has revolutionized various aspects of modern life, and it's no secret that this transformation comes with a significant demand for energy. Understanding and managing energy consumption in AI systems, particularly smart assistants, is critical in aligning technological advancements with sustainable practices. This section delves into strategies and innovations aimed at minimizing energy usage while maximizing efficiency.

Technology's widespread integration into everyday tasks is largely supported by vast networks of servers and data centers. These infrastructures are responsible for processing the immense amounts of data that AI systems, including smart assistants, rely on. Consequently, they consume vast quantities of electricity, contributing to their environmental footprint. But this isn't just about the energy itself; it's about the impact of energy sources, which vary widely in terms of carbon emissions. To mitigate the environmental consequences, companies are turning their attention to renewable energy solutions and innovative technologies to power these servers sustainably.

One of the primary strategies involves optimizing algorithms to reduce computational load. An algorithm that requires fewer processing cycles can dramatically decrease energy consumption, leading to more efficient AI systems. Developers focus on refining these algorithms, utilizing techniques like pruning neural networks, which effectively strips away unnecessary parts of the AI's learning framework to streamline operations. Each improvement in efficiency means a reduction in the energy needed to perform tasks, thus contributing to less energy-intensive operations.

Another exciting avenue is the development of AI-specific hardware aimed at energy conservation. Traditional CPUs and GPUs are versatile but not necessarily efficient for all AI tasks. Enter dedicated AI processors and chips—designed to execute AI computations more efficiently, these components provide a means to manage energy consumption. These hardware innovations offer specialized architectures that speed up data processing while simultaneously lowering power requirements.

Beyond technical enhancements, companies are also rethinking their data center strategies. Data centers optimized for energy consumption can play a pivotal role in reducing the overall environmental impact of AI technologies. Techniques like advanced cooling systems and passive cooling strategies are being implemented to manage heat efficiently. By maintaining optimal temperatures through eco-friendly methods, these facilities consume less energy, enhancing the sustainability of the digital backbone that supports smart assistants.

Virtualization and cloud computing present another opportunity for energy management. By utilizing the cloud, which benefits from economies of scale and enhanced resource allocation, the energy footprint of smart assistants can be minimized. Cloud providers often invest in state-of-the-art energy-efficient technologies, which individual users indirectly access through their smart devices. This shared infrastructure reduces the need for individual high-energy-consuming hardware, lowering the environmental burden associated with personal AI use.

Moreover, energy management isn't just a technological challenge—it's a strategic concern. Businesses have shifted towards holistic approaches that examine energy use across the entirety of their operations. Energy audits and assessments help organizations pinpoint areas of excessive consumption, allowing them to implement specific solutions tailored to their needs. By integrating energy management into business strategies, companies not only reduce their carbon footprints but also potentially lower operational costs.

There's also a growing emphasis on transparency in energy usage. By providing detailed reports on energy consumption and associated emissions, tech companies can drive accountability and encourage more sustainable practices industry-wide. These reports can influence consumer behavior, as users increasingly seek out services and products from companies that demonstrate a commitment to environmental responsibility. Through transparency, businesses can guide industry standards and inspire both competition and collaboration towards better energy management practices.

Public awareness and policies are also pivotal in this discussion. As awareness of AI's energy use grows, so too does demand for governmental regulations that push for energy-efficient innovations. Policy frameworks can incentivize organizations to invest in green technologies. By setting industry benchmarks for energy consumption, policy guidance can steer tech development towards choices that emphasize sustainability. Public-private partnerships may spur further progress, facilitating the development and adoption of energy-efficient technologies.

While the task of managing energy consumption in AI seems daunting, it's essential to view this challenge as an opportunity. Innovative solutions have the potential to not only improve energy efficiency but also to redefine how we interact with technology in an environmentally-conscious manner. As consumers and businesses become more aware of the environmental impacts, continuous improvement and adaptation to greener practices gain momentum. The goal is to harmonize technological advancement with sustainable practices, ensuring that the promise of AI doesn't come at the planet's expense.

Looking forward, there's hope and optimism as new solutions continually emerge. The convergence of AI technology with sustainable practices signals an era where smart assistants become not just tools of convenience but also harbingers of a cleaner, more energy-efficient future. As we stand on the cusp of further AI evolution, the pursuit of energy consumption management becomes not just a necessity but a defining cornerstone of technological innovation. By prioritizing energy-efficient AI development, we can transform the ways smart assistants influence our lives, crafting a future where technology supports not only productivity and convenience but also planetary stewardship.

AI in Sustainable Practices

The intersection of artificial intelligence and sustainability presents a compelling frontier in contemporary technological discourse. AI has the potential to drastically redefine sustainable practices in ways that were previously unimaginable. By leveraging sophisticated algorithms and real-time data, AI systems can optimize resources, develop sustainable energy solutions, and even forecast environmental changes with unpreceded accuracy.

One of the most profound impacts AI can have on sustainability is through energy consumption management. Traditional methods of energy management often rely on historical data, whereas AI can continuously analyze patterns and predict future consumption needs, allowing for dynamic adjustments. This ensures energy is used more efficiently, reducing waste and diminishing the carbon footprint of entire nations.

AIs like smart assistants play a crucial role in sustainable lifestyle choices by recommending energy-efficient options and providing insights on reducing personal consumption. Whether it's suggesting times for off-peak energy use or controlling smart homes' thermostats and lights, these assistants make conserving energy seamlessly integrated with daily routines. The ability of AI to regulate appliances' energy use in real-time means homes and businesses can operate more sustainably.

In agriculture, AI is revolutionizing sustainable practices through precision farming. With AI-driven insights, farmers can make informed decisions about when and where to plant crops, how much water to use, and the optimal times to harvest. This precision in farming ensures resources are used efficiently, reducing waste and increasing crop yields—a necessity for feeding the growing global population sustainably.

AI also enhances supply chain efficiency, a critical component of sustainable practices. By analyzing logistical data, AI can streamline operations, reduce fuel consumption, and minimize the environmental impact of transporting goods. Smart assistants help businesses track their supply chains, forecast demand, and adjust routes to ensure the least environmental damage possible—a small step towards a broader concept of green logistics.

Urban planning and development are another vital area where AI makes headway. AI helps design smart cities that prioritize sustainability through reduced energy consumption and improved waste management. It can simulate different city layouts, assess their environmental impact, and suggest the most efficient designs. By collaborating with city planners, AI ensures new urban developments are both technologically advanced and environmentally friendly.

The impact of AI in managing human-wildlife conflict can't be overlooked. AI technologies help monitor endangered species and their habitats by analyzing data collected from drones and cameras. This aids in protecting biodiversity, a key element in preserving ecosystems for future generations.

Moreover, AI assists in climate modeling—a complex task requiring the analysis of vast datasets. Advanced machine learning algorithms forecast environmental trends, informing policies aimed at combating climate change. These predictions enable precise planning for disaster response and mitigation, ultimately reducing the potential damage and building resilient communities.

While AI has immense potential, it's important to address its own environmental impact. Training AI models requires significant computational power, which in turn consumes a considerable amount of energy. Thus, striking a balance between deploying AI technologies and minimizing their ecological footprint is crucial. Efforts towards developing more energy-efficient AI models are underway, such as optimizing algorithms to require less data processing power.

Collaboration between tech companies, governments, and environmental organizations is essential to harness AI's potential sustainably. Policies can guide AI's development and deployment, ensuring its use aligns with broader environmental goals. Initiatives such as carbon pricing and sustainable procurement policies could drive companies to adopt AI technologies that prioritize environmental preservation.

AI-driven sustainable practices offer hope in the fight against environmental challenges. As this technology continues to evolve, its integration into sustainable practices will likely grow more sophisticated and influential. The journey towards a greener world is ongoing, and with AI in the driver's seat, humanity can look forward to innovative solutions tackling climate challenges.

Chapter 20:
Cultural and Ethical Considerations

As smart assistants weave themselves intricately into diverse facets of our lives, understanding their cultural and ethical resonances becomes essential. This technological tapestry, while promising enhanced ease and productivity, often navigates complex cultural terrains where values and norms differ widely. AI can inadvertently reflect and amplify biases if not meticulously calibrated to honor non-homogeneous cultural contexts. Ethical implications, too, demand vigilant contemplation, as AI's decision-making power hikes into realms traditionally governed by human judgment. Will our reliance on these digital intelligences erode the moral accountability traditionally innate to human choices? As we venture deeper into this AI-powered era, these considerations beckon us to cultivate systems that not only augment human capabilities but do so with a conscience tuned to society's ethically diverse harmonies.

AI Across Different Cultures

The pervasive influence of AI in modern life isn't confined by geographical boundaries. As smart assistants proliferate globally, they reflect and adapt to diverse cultural contexts, leaving traces of their origins and adapting to local nuances. Just as language and customs vary from one region to another, so do the expectations and interactions with artificial intelligence. This section delves into the fascinating interplay between AI technology and cultural diversity, examining how AI assistants are understood, utilized, and perceived within distinct societal frameworks.

In many Western countries, AI is often greeted with excitement and curiosity, embraced as an harbinger of efficiency and productivity. Here, individuals are inclined to integrate these digital helpers into various facets of everyday life—work, education, and entertainment. People in such cultural settings tend to prioritize convenience, aspiring to leverage AI's capabilities to streamline actions, sort through overflowing schedules, and optimize day-to-day operations. The narrative often circles around empowerment, highlighting how AI can liberate users from mundane tasks and free up time for more meaningful pursuits.

Conversely, in other parts of the world, AI might be met with skepticism or caution. Concerns about privacy, data security, and the implications of widespread automation echo loudly in cultures where people are more reserved about adopting new technologies. For instance, in regions with a strong focus on community and interpersonal relationships, there can be a prevailing apprehension regarding AI's role in replacing human jobs and dehumanizing personal experiences. As a result, the adoption rate might be slower, but the ethical discourse is more pronounced, encouraging developers and policymakers to consider local values and priorities when introducing AI solutions.

In many Asian countries, like Japan and South Korea, there exists a unique harmony between cutting-edge technology and age-old traditions. Here, smart assistants might seamlessly blend into the cultural landscape, simultaneously fostering innovative practices and respecting traditional values. Such cultures often have a strong inclination towards technology, driven by both fascination and necessity. The integration of AI in these countries isn't merely about facilitating tasks; it can also act as a bridge connecting youthful ingenuity and ancient wisdom. This duality fosters the development of AI assistants that not only understand commands but are also culturally attuned to recognize the subtleties of social etiquette and linguistic nuances.

In Africa, the narrative around AI is one of potential and opportunity. As smart assistants extend their reach into this diverse continent, they are viewed as catalysts for development. In regions where infrastructure might be challenging, AI can play a pivotal role in bridging gaps in education, healthcare, and commerce. The cultural embrace here often involves a practical outlook, where AI is seen not only as an assistant but as an equalizing force that democratizes access to resources and knowledge. Developers in these regions might focus on localization, ensuring that AI applications are available in various dialects and tailored to solve region-specific challenges such as agricultural planning or resource management.

The Middle East offers another perspective, where AI technology is analyzed through the lens of tradition and modernity. Here, technological advancements are often pursued with vigor, but always with cultural coherence in mind. AI assistants find themselves navigating a landscape where sensitivity to cultural norms, religious beliefs, and historical contexts is crucial. In such environments, success relies on the ability to provide practical solutions while respecting conservative values—an intricate balancing act that demands sophistication and insight from AI developers.

In Latin America, AI's growing presence is shaped by vibrant cultures and social complexities. The integration of smart assistants into daily life reflects a dynamic blend of technological enthusiasm and sociopolitical considerations. As these nations grapple with issues like economic disparity and digital access, AI emerges as a double-edged sword—harboring the promise of economic growth and modernization but also raising questions about equity and inclusion.

Here, the discourse often centers on ensuring that AI technologies serve the broader community and do not exacerbate existing inequalities. As such, initiatives aimed at fostering digital literacy and inclusion tend to be at the forefront of AI development strategies.

The beauty of AI's cultural journey is that it's inherently malleable. Unlike static technologies, AI has the unique ability to learn and adapt. This means that as it traverses different cultures, AI has the potential to evolve, reflecting the values, traditions, and social norms of its users. Whether through language processing that captures regional dialects or through algorithms that customize services to suit local preferences, AI can transcend its origins to resonate deeply with the people it aims to assist.

Nevertheless, as AI becomes more culturally aware, there is an undeniable need for further exploration of ethical implications. Each interaction between culture and AI poses questions about autonomy, representation, and bias. Developers must be conscientious, avoiding the imposition of a singular narrative that could overshadow local realities. Instead, a collaborative approach that involves local stakeholders can ensure that AI technologies contribute positively and meaningfully within various cultural landscapes.

The journey of AI across different cultures presents both challenges and opportunities. It is a testament to human ingenuity that a digital construct originally conceived in the context of one culture can adapt and thrive in a multitude of different environments. As we look to a future where AI is even more ingrained in the fabric of daily life, the relationship between technology and culture will likely become even more complex and richly woven, offering endless possibilities for innovation while demanding a continual dialogue about its cultural and ethical dimensions.

Ethical Implications of AI Usage

In embracing the convenience of AI-powered smart assistants, particularly when they seamlessly integrate into our daily lives, we must not overlook the profound ethical implications that come with their usage. As these digital companions become increasingly embedded in aspects of our personal and professional activities, they hold the potential to reshape not just how we do things, but also how we perceive fundamental ethical principles like privacy, accountability, and equity.

The allure of smart assistants lies in their capability to anticipate our needs, streamline tasks, and offer tailored recommendations. However, this very prowess opens a Pandora's box of ethical dilemmas. One primary concern revolves around *data privacy*. To function effectively, these assistants require access to vast amounts of personal data. From listening patterns to location tracking, they continuously gather user information to enhance functionality. While this data collection cultivates an increasingly personalized experience, it also raises alarms about how such sensitive information might be used or misused.

The potential for privacy breaches is not hypothetical. Instances of data mishandling have occurred, causing distrust among users. As we look forward, the challenge lies in striking a balance between data utility and protection. Companies developing smart assistants must ensure that they incorporate robust encryption methods, transparent data policies, and give users more control over their information. Only then can trust be fostered in the digital landscape—and this trust is a cornerstone of ethical AI usage.

Algorithmic Bias, another ethical challenge, deserves our attention. AI systems, including smart assistants, are often perceived as impartial decision-makers. However, they function based on the datasets they are trained on. If these datasets reflect certain biases, the

AI is likely to perpetuate those biases, sometimes with far-reaching consequences. For instance, facial recognition technologies have been criticized for inaccurate results across different ethnicities, not because the technology itself is inherently biased, but because the data used to train it was not suitably diverse.

Addressing algorithmic bias requires a multi-faceted approach. Developers need to prioritize diversity and inclusivity in their training data and involve ethicists at every stage of AI development. Furthermore, ongoing testing and evaluation are crucial to identify and mitigate bias in real-time. These steps not only promote fairness but also enhance the overall reliability and acceptance of AI systems.

In parallel with privacy and bias, the ethical implications of AI on employment cannot be ignored. As smart assistants continue to automate functions previously done by humans, there's an ever-growing concern about the displacement of jobs. This shift demands a forward-thinking approach to workforce development. By investing in education and upskilling programs, societies can prepare individuals for the evolving job landscape, ensuring that technology acts as a complement to human capabilities rather than a replacement.

Beyond individual implications, AI usage has broader societal impacts. The way in which AI is integrated into our societal frameworks can either bridge or exacerbate social inequalities. On the one hand, AI has the power to democratize access to resources, such as healthcare and education, helping underserved communities. On the other hand, the digital divide—the gap between those who have easy access to digital technology and those who do not—can deepen if AI tools increasingly become the norm without equal access.

The ethical journey of AI usage also intersects with *accountability*. If a smart assistant makes a mistake leading to a negative outcome, who is to blame? The developer? The user? The technology itself? Current legal frameworks often lag behind technological realities, struggling to delineate clear lines of accountability. Navigating this ethical terrain requires new regulations that keep pace with the rapid development of AI technologies, clearly defining what constitutes responsible and ethical behavior in AI systems.

Transparency is a key component in fostering accountability. Users should be aware of how decisions are made, what data is used, and who has access to their information. Implementing AI systems in an open manner not only helps to build trust but also empowers individuals to make informed choices about their technology use.

Last but not least, ethical AI usage demands a visionary approach where the focus extends beyond the immediate implications of AI deployment. It requires envisioning futures where AI-driven societies prioritize human dignity, equity, and well-being. Those involved in AI development and regulation must hold onto long-term ethical considerations even as they address current technological challenges.

In conclusion, as AI continues to redefine realms of possibility, it brings with it the responsibility to ensure ethical governance. By proactively addressing the ethical implications of AI usage, we can steer the impact of smart assistants in a direction that benefits humanity as a whole. The choices we make today in terms of policy, design, and education will sculpt the digital landscape of tomorrow, ensuring that technology remains a force for good, aligned with our shared values and ethical principles.

Chapter 21:
Legal Aspects of Using AI Assistants

As AI assistants increasingly weave themselves into the fabric of our daily routines, understanding the legal landscape surrounding their use becomes crucial. With data privacy concerns at the forefront, user agreements often serve as the fine print in our technological interactions, outlining the rights and responsibilities of both parties. But the legal terrain extends beyond privacy. Intellectual property issues arise too, as AI-crafted content challenges traditional ownership definitions, demanding new frameworks for recognition and protection. While AI offers unprecedented conveniences, users need to navigate these agreements carefully, aligning technological aspirations with legal obligations, and ensuring a harmonious digital future that respects both innovation and regulation.

Understanding User Agreements

Navigating the legal labyrinth that surrounds AI assistants begins with understanding user agreements. These documents, often dismissed as mere formalities, are crucial contracts that outline the responsibilities between users and providers. They are not just lengthy texts filled with legalese; they play a fundamental role in defining how users interact with smart technologies.

User agreements are the bridges connecting legal stipulations to everyday user experiences. They cover a wide range of issues, including user data handling, privacy implications, and terms of service. By

139

agreeing to these terms, users enter into a contractual relationship with the service provider, accepting the rules set forth within.

One of the pivotal aspects of user agreements is data privacy. Smart assistants rely heavily on user data to provide personalized services. The user agreement outlines how this data is collected, stored, and shared. Users should be aware of how their data will be utilized and the potential risks involved. It's a delicate balance between convenience and privacy, often leaning toward the former in the interest of enhanced service.

Another significant area covered by user agreements is liability. These legal documents define the extent to which the service provider is responsible for any issues that may arise from the use of the AI assistant. From device malfunctions to data breaches, the agreement stipulates the legal protections the provider extends — or limits — to its users.

For instance, user agreements often include clauses that limit the provider's liability in the event of service interruptions or technical failures. This stipulation highlights the importance of understanding what users are agreeing to and ensuring they are prepared for any potential risks or limitations inherent to the technology.

These agreements also delve into the acceptable use of the technology. They outline prohibitions against behavior that could be deemed damaging or illegal, ensuring that users are aware of the ethical and legal boundaries that accompany the use of AI assistants. Such clauses are crucial for maintaining ethical use and adherence to legal standards in various jurisdictions.

Moreover, user agreements must be dynamic, reflecting the changing landscape of technology and law. As AI evolves, so too do the legal frameworks that govern its use. Service providers are obligated to update agreements to comply with new regulations and advancements in AI technology, providing users with the latest contractual terms and protections.

User agreements are often amended to address emerging issues such as algorithmic bias or to enhance transparency about how AI decisions are made. Regular updates ensure that users are not only kept informed but also legally protected in an ever-evolving digital world.

For users, understanding user agreements requires a balance between legal literacy and pragmatic engagement. While the technical jargon can be daunting, it's essential for users to comprehend the implications of these agreements thoroughly. Reading through the terms, even if it seems tedious, empowers users to make informed decisions about their digital interactions.

Furthermore, the role of user agreements in fostering trust should not be underestimated. When users are confident that their data is handled responsibly and that the service provider is committed to ethical practices, it enhances the trust relationship, which is paramount for the widespread adoption of AI technologies.

Developing a habit of scrutinizing these agreements can facilitate a more comprehensive understanding of the potential impacts of technology on daily life. It's essential for users to be proactive in questioning provisions that seem unclear or concerning, seeking clarifications where necessary.

In summary, user agreements are indispensable components of the AI assistant ecosystem. They safeguard the interests of both users and providers, ensuring that the latter can innovate and offer services while respecting privacy and legal obligations. By understanding these agreements, users can harness the potential of smart assistants while remaining vigilant about their rights and responsibilities.

In conclusion, embracing AI technologies responsibly starts with acknowledging the value of user agreements. They are the foundation of a legal relationship between users and providers, guiding the ethical and secure use of AI assistants. As tech-savvy individuals venturing into a world defined by smart technology, understanding these documents empowers us to fully utilize AI assistants while safeguarding our interests in an increasingly connected world.

Intellectual Property Issues

Incorporating AI assistants into daily life doesn't just add convenience and novelty; it also invites a plethora of legal considerations, chiefly among them, intellectual property (IP) issues. The development and deployment of these intelligent tools rest squarely in the realm of creativity and innovation, birthing new dimensions of IP law. As AI continues to evolve, it relentlessly tests the boundaries of existing IP frameworks, challenging creators, developers, and legal professionals to keep pace.

Intellectual property law is designed to protect the rights of creators and inventors, encompassing patents, copyrights, and trademarks. With AI assistants, the lines blur when determining ownership and originality. You might wonder how an AI's ability to generate content—like music, art, or even written work—fits into this equation. Who holds the rights to AI-generated content? Is it the user who prompted the creation, the developer of the AI, or the AI itself? Legal systems worldwide are still wrestling with these questions.

Consider copyright law, which traditionally shields the expression of original ideas. The ambiguity arises as AI-generated works—often indistinguishable from human creations—emerge. Current laws generally dictate that copyright protection requires human authorship. Therefore, when an AI generates a novel piece of art, the questions of ownership become more convoluted. While some argue that the input of the human operating the AI should grant them ownership rights, others see the software developers as rightful owners, having crafted the system that enables such creativity.

Then we delve into the realm of patents, where technology advancements push the envelope even further. AI assistants entail numerous patented technologies, from voice recognition algorithms to data processing techniques. This landscape becomes even more complex when AI systems themselves begin innovating new methods or technologies. If an AI invents something entirely novel—a process, a solution, a product—does the patent belong to the inventors of the AI, to the machine itself, or perhaps to the entity using the AI? Most jurisdictions don't recognize non-human inventors, but legislative winds are shifting amid unprecedented developments.

Trademark issues come to the fore in scenarios where AI assistants interact with brand identifiers. As these smart systems engage in e-commerce, advertising, or provide customer support, the potential misuse or misrepresentation of trademarks could lead to disputes. If an AI assistant inadvertently misuses a brand name or slogan, causing consumer confusion or diluting a brand's identity, questions about liability arise. Who is responsible—the user, the company deploying the AI, or the creators of the technology?

Moreover, AI assistants often rely on vast datasets to function effectively. Among these data are copyrighted materials—texts, music, images—that aid machine learning models. The use of this data without clear rights presents another IP conundrum. In jurisdictions with fair use policies, using snippets of copyrighted material for AI training might not face legal battles, but elsewhere, the boundaries are not as distinct. As AI technology grows, the usage rights of training data remain under intense scrutiny, requiring careful navigation to avoid potential infringements.

Collaboration between legal experts and tech developers is crucial in addressing these evolving IP issues. Patents may require updates to recognize AI's contributions, while copyright laws might need amendments to include ownership criteria for AI-generated works. Typically, policy reform moves at a glacial pace compared to technological advancement. Hence, preemptive measures such as AI usage guidelines, and agreements among stakeholders can mitigate potential conflicts.

The future promises further implications of AI on IP law. Consider autonomous AI: as these systems become more advanced, the notion of attributing them with pseudo-personhood might surface. While granting an AI legal standing is speculative, it's a concept under serious academic and legal discourse. For now, the prevailing thought remains that humans—whether AI developers, operators, or businesses—are accountable for an AI's creations and associated IP rights.

Businesses using AI assistants should develop robust strategies for managing potential IP issues. This includes clarifying ownership rights in contracts with AI developers and being diligent about IP audits in their digital activities. Protecting proprietary technologies and original content while respecting others' IP rights is crucial for maintaining competitive advantage and avoiding costly legal disputes.

Ultimately, the intersection of AI and intellectual property challenges traditional notions of authorship, ownership, and innovation. As society navigates this digital transformation, the legal landscape must adapt to ensure that creativity and technological advancement continue to thrive without undermining the foundational principles of intellectual property law. Tackling these issues head-on will enable AI to flourish as an augmentative tool in modern life, enhancing productivity while safeguarding the rights and interests of all stakeholders involved.

Chapter 22:
Community and Social Interaction

In an era where digital connections define our social landscapes, AI-powered smart assistants have emerged as instrumental tools in fostering community and social interaction. They break down communication barriers by streamlining the management of social media platforms, transforming how we engage with our networks and allowing for more fluid exchanges of ideas and emotions. This marriage of technology and connectivity empowers individuals to build vibrant online communities that transcend geographical constraints, enriching our social experiences while encouraging collective growth. By harnessing AI's capability to analyze and respond to social cues, these assistants contribute not only to personal efficiency but to the creation of socially dynamic spaces where people can collaborate, share, and innovate like never before. As we continue to integrate these intelligent systems into our social spheres, they challenge us to rethink the fabric of community itself, urging us toward more inclusive and meaningful interactions in the digital age.

AI in Social Media Management

In today's digital ecosystem, social media platforms have become a vibrant tapestry of community and interaction, weaving together narratives from across the globe. At the heart of this evolution, artificial intelligence, particularly smart assistants, are now playing a pivotal role in the intricate dance of social media management. As we delve deeper into AI's role here, one can't help but appreciate the blend of technology and social dynamism it ushers in.

AI in social media management serves as an extraordinary ally, streamlining tasks that were once tedious and time-consuming. From scheduling posts to analyzing audience engagement, AI tools offer precision, efficiency, and insights that human managers might overlook amidst the online noise. The ability to auto-schedule content ensures that the intended message reaches audiences at optimal times, stretching across different time zones, without disruption. This proactive engagement keeps communities vibrant and consistently interactive.

The magic doesn't stop there. AI-driven analytics provide a real-time pulse check on audience sentiment, dissecting variables like likes, shares, comments, and non-verbal interactions. With these insights, businesses and individuals can tailor their approach, crafting content that resonates and drives meaningful interaction. This adaptability is crucial for building and sustaining communities that are not only active but also deeply engaged. AI assists in recognizing patterns and trends that could otherwise take a human eye much longer to detect, enabling more strategic decision-making.

Another significant contribution of AI to social media management is in handling the vast ocean of user-generated content. Platforms often grapple with how best to moderate content while maintaining the freedom of expression essential for any community. AI algorithms, trained to identify harmful or inappropriate content, act as vigilant guards, ensuring that platforms remain safe and welcoming spaces. This allows community managers to focus on fostering positive interactions and collaborations among users.

However, AI's role in this space transcends just management and moderation. It's reshaping how brands build their identity and narrative, allowing for more personalized experiences. With AI, content can be dynamically tailored to each user's preferences and interests, enhancing the relevancy and impact of interactions. This personalization creates a sense of inclusion, making community members feel as though their unique tastes and opinions are acknowledged and valued, thereby strengthening community ties.

The aspect of customer service on social media platforms is another field being redefined by AI. Through chatbots and virtual assistants, companies can provide immediate answers to user queries, ensuring that the customer experience is seamless. These AI tools learn and evolve with every interaction, improving their accuracy and response quality over time. By handling customer service efficiently, they free up human resources for more complex problems that require empathy and nuanced understanding.

It's worth noting that smart assistants are not just tools for large corporations with vast online presences. Small businesses and individual content creators can also harness their capabilities to maximize reach and growth. Advanced targeting features, insights, and AI-powered creative tools democratize access to sophisticated marketing strategies that were once the domain of industry giants. This accessibility is crucial in leveling the playing field, allowing a diverse range of voices and ideas to flourish.

Yet, we'd be remiss if we didn't acknowledge the challenges that accompany AI in social media management. There are concerns about privacy, data security, and the potential for AI to perpetuate biases. Responsible use of AI requires a careful balance between harnessing its power for community building and being mindful of ethical and regulatory boundaries. Social media platforms and AI developers must collaborate closely, enhancing transparency and trust within digital communities.

Despite these challenges, the potential of AI in reshaping social media landscapes is undeniable. It can help us move toward a more connected, informed, and interactive digital world. As AI continues to evolve, its partnership with human creativity and judgment will be crucial in crafting experiences that are not only efficient but also authentic and inspiring. Within this rapidly changing environment, those who learn to leverage AI smartly will undoubtedly carve out spaces rich in interaction and opportunity.

As we stand on the brink of further innovations, the fusion of AI and social media begs a fundamental question: How do we wish to engage and connect with others in this digital age? By embracing AI's potential thoughtfully and strategically, we open doors to richer social interactions that echo the core values of collaboration, understanding, and growth that first drew us to these platforms.

Building Online Communities

In today's interconnected world, smart assistants are not only transforming individual experiences but also empowering the formation and management of thriving online communities. As tech-savvy individuals seek to enhance their digital lives, the integration of AI-powered tools into community building provides unique opportunities for engagement, collaboration, and support. The virtual platforms where these communities thrive are dynamic spaces that evolve with technological advancements, particularly those centered around AI.

Online communities have long been a cornerstone of digital interaction, connecting individuals with shared interests and goals. The introduction of AI into these spaces offers a level of personalization and interactivity previously unattainable. Smart assistants can facilitate seamless communication by managing routine interactions, automating responses, and filtering content to match the interests of community members. This functionality allows for a more focused and tailored community experience, where members receive content that resonates most with their preferences.

One of the key benefits AI brings to online communities is enhanced engagement. Smart assistants can identify patterns in member behavior and interaction, offering insights into what drives community participation. This data can be used by community moderators to refine their strategies, creating content and events that boost engagement. A dynamic interaction approach helps sustain membership and fosters a vibrant and active community, fostering a sense of belonging among members.

Moreover, AI-powered smart assistants reduce the barrier to entry for those looking to start new communities. With AI's capacity to handle administrative tasks, such as approving new members or moderating content for compliance with community guidelines, founders can focus more on strategic community growth. This democratization of community management allows more individuals to spearhead initiatives without the need for exhaustive resources traditionally required in community-building efforts.

Smart assistants also play an essential role in maintaining secure and respectful environments within these online ecosystems. AI-driven moderation tools can automatically flag or remove inappropriate content, ensuring that communities remain civil and inclusive. This real-time moderation capability not only protects members but also upholds the integrity of the community, encouraging greater trust and participation among users.

The integration of AI in community building extends beyond mere administration and security. It also enhances the educational potential of these platforms. Communities centered around learning and knowledge-sharing greatly benefit from AI, as smart assistants can curate educational content, suggest learning paths, and even facilitate peer-to-peer knowledge exchange. This form of assisted learning encourages members to develop their skills collaboratively, creating an environment of mutual growth and support.

Of course, the widespread use of AI in community building raises pertinent questions about privacy and data security. As smart assistants gather data from user interactions, there is a growing need to ensure that this information is handled ethically and securely. Community leaders must be transparent about how data is used and stored, instilling confidence within their communities by implementing robust data protection measures. Clear guidelines and best practices surrounding data security must be established to safeguard members' privacy and maintain trust.

AI's ability to manage and analyze large volumes of information also offers new insights into community dynamics and trends. Communities can leverage these insights to adapt to changing member needs and expectations, facilitating a flexible approach to growth and engagement strategies. This adaptability ensures that online communities remain relevant and valuable to their members, empowering them to evolve in tandem with technological and social innovations.

The future of online communities supported by AI looks promising and expansive. As smart assistants become more sophisticated, their role in community building will only deepen. They will likely become intuitive partners capable of predicting and fulfilling member needs, enhancing the overall community experience. Such advancements could lead to a new era of intelligent communities where members are both contributors and beneficiaries of collective wisdom, driven by AI insights.

In conclusion, smart assistants are reshaping the landscape of online community building. By enhancing engagement, simplifying management, and fostering meaningful interactions, AI is laying the groundwork for more inclusive and dynamic virtual communities. As we look to the future, these developments invite us to imagine a world where our digital interactions are as rich and fulfilling as our real-world connections, propelled forward by the synergistic relationship between human initiative and AI support.

Chapter 23:
AI in Crisis Management

In a world increasingly fraught with uncertainties, AI stands as a beacon of hope in crisis management, offering unprecedented capabilities to anticipate, prepare for, and respond to emergencies. Harnessing vast amounts of data, smart assistants can analyze real-time information to predict potential disasters, enabling proactive measures that can save lives and minimize damage. From coordinating emergency response logistics to providing real-time updates during hurricanes and earthquakes, AI-driven systems act as invaluable tools for first responders, enhancing their effectiveness and efficiency. These intelligent solutions empower communities by equipping them with timely alerts and actionable insights, facilitating better decision-making during critical moments. As AI continues to evolve, its role in crisis management will expand, offering even more innovative ways to manage and mitigate the impacts of global events, ultimately fostering a safer and more resilient future for all.

Emergency Preparedness and Response

In a world that's increasingly fraught with unpredictability, the role of artificial intelligence in crisis management cannot be overstated. Advances in AI technology have revolutionized how we approach emergency preparedness and response strategies, offering unprecedented opportunities for enhancing safety and efficiency during critical situations. AI has the potential to not only anticipate crises but also to effectively manage and mitigate their impact.

Emergency preparedness is all about getting ready before disaster strikes. AI systems can analyze massive datasets to predict potential hazards and their likely impact. For instance, AI algorithms can process weather patterns and geological data to forecast natural disasters such as hurricanes or earthquakes weeks in advance, giving communities precious time to prepare. With this kind of warning, evacuation plans can be enacted, resources can be allocated, and chaos can be minimized.

A critical component of emergency preparedness is communication. AI-powered smart assistants are transforming how we disseminate alerts and updates. They can rapidly send out notifications over multiple platforms, ensuring that information reaches people when it matters most. Whether it's a voice-activated device in your home coaxing you to seek shelter or a multi-channel alert sent to your phone, AI assists in reducing the communication gap during crises.

In addition to prediction and communication, AI plays a pivotal role in data management during emergencies. The influx of information during crises can be overwhelming. AI systems help sift through data in real-time, prioritizing crucial information to aid decision-makers in formulating informed responses. This efficiency is vital when every second counts, helping emergency services to deploy resources where they're most needed.

An intelligent response to emergencies doesn't end with crisis anticipation. AI's capability extends to a more robust response mechanism that enhances traditional methods. Drones powered by AI are now on the front lines, collecting aerial data to assess damage and locate individuals needing assistance. They're particularly beneficial in situations where it's too risky for humans to venture. The data collected is instantaneously analyzed to coordinate rescue missions and supply distribution with precision.

AI-powered facial recognition and biometric systems are also aiding in identifying missing persons and reuniting families. These systems can scan through numerous images and features far quicker than humans, providing hope in dire circumstances. Additionally, AI-driven applications can support mental health by offering customized support to individuals coping with trauma post-crisis, making resilience and recovery possible.

Machine learning, a subset of AI, allows systems to adapt and improve after each emergency. By analyzing their performance and learning from the outcomes, these systems continuously enhance their response strategies. It's a matter of turning experience into expertise, ensuring that systems evolve to meet new and unpredictable challenges. In essence, AI becomes better at managing crises the more it's used, providing a reliable foundation for future emergency responses.

AI's contribution to crisis management isn't just about enhancing tools—it's about supporting humans. By shouldering the burden of data processing and repetitive tasks, AI allows human responders to focus on tasks requiring empathy, judgment, and human intuition—elements machines cannot replicate. The symbiotic relationship between humans and technology in crisis settings represents a harmonious blend of strengths, producing outcomes that neither could achieve in isolation.

The hurdles in implementing AI for emergency management widely stem from the availability of infrastructures, data privacy concerns, and technical limitations. Yet, these challenges present opportunities for growth and innovation in the AI sector. By investing in robust infrastructure and conducting thorough data security audits, we can overcome these barriers and pave the way for a secure AI-driven future.

Community engagement and public awareness remain crucial in fostering trust in AI systems. People need to understand the role of AI in safety and should be educated about its capabilities and limitations. Transparent communication between tech developers, emergency services, and the general public fosters cooperation and understanding, potentially improving the adoption and effectiveness of these systems.

AI in crisis management isn't science fiction anymore—it's a practical and powerful reality. By acknowledging its potential and supporting continuous development, society can be better prepared for the unpredictable. With AI at our disposal, emergency preparedness and response become proactive and anticipatory rather than reactive, ensuring a safer and more secure future for all.

Supporting Disaster Relief Efforts

When disaster strikes, the chaos and urgency can overwhelm traditional relief efforts. This is where AI, specifically through smart assistants, steps up as a formidable ally. Advanced AI technologies are reshaping the landscape, offering unmatched support to first responders and humanitarian aid organizations through precise data analytics and rapid information dissemination.

Smart assistants are proving invaluable in disaster scenarios by acting as immediate information hubs. They gather data from multiple sources—social media, news feeds, weather updates—and process it in real-time. By doing so, they help pinpoint affected areas, assess the severity of damage, and even predict the trajectory of natural disasters. This critical function can be lifesaving, enabling teams to allocate resources efficiently and make informed decisions swiftly.

Imagine a situation where a hurricane looms on the horizon. Sensor networks and satellite imagery, analyzed by AI, feed into smart assistants. These assistants can then alert nearby populations, offer safety instructions, and keep the public updated on evacuation routes or shelter availability. And when communication lines are down, AI-powered systems with offline capabilities can still provide crucial guidance locally, ensuring that no one is left in the dark.

Beyond prediction and immediate response, smart assistants also play a crucial role in post-disaster recovery. By analyzing disaster impact data, these systems can help in strategizing long-term rehabilitation plans. For instance, they can assist in assessing which infrastructures need urgent attention, prioritize resource allocation, and monitor the progress of recovery efforts over time. With enhanced analytical capabilities, AI helps not only to rebuild but to build back better, fostering resilience against future calamities.

Collaboration between AI developers and humanitarian organizations is vital to maximize the use of smart assistants in disaster relief. Engaging with local communities and understanding their needs ensures that the technology is effectively tailored to provide genuine value. Encouraging communication between tech innovators and response teams leads to the development of intuitive systems that can anticipate and adapt to various scenarios on the ground.

Another area where AI comes to the forefront is in documenting and assessing the impact of disasters for future reference. By creating detailed reports using machine learning and data analytics, smart assistants can help in understanding patterns and trends. This learning aids in formulating better preparedness plans, adapting equipment and methods, and mitigating risks more effectively in subsequent events.

Smart assistants, powered by AI, bring a level of precision and efficiency previously unattainable in emergency situations. They transform raw data into actionable insights, making complex procedures more manageable for human teams. However, it's important to remember that while these assistants provide support, the empathy and judgment that human responders bring to the table remain irreplaceable components of disaster relief efforts. Technology complements human intuition, rather than replacing it.

The ability of AI to learn and improve over time ensures that the support it offers becomes increasingly targeted and effective. As algorithms become more advanced, they will be able to predict complex emergency management dynamics, prepare automated simulations to test different response strategies, and manage vast logistical operations with remarkable agility.

Moreover, for those in tech who are passionate about contributing to society, disaster relief offers a field where innovation can make a tangible difference. Developers have the opportunity to create scalable AI solutions that transform how we respond to humanitarian crises, making each innovation a step toward saving more lives and reducing suffering.

While the focus has often been on improving efficiency and productivity in everyday settings, when it comes to crisis management, AI's capability extends far beyond convenience. It becomes a mission-critical tool that aids humanity in its hardest moments, carving paths for future advancements that could redefine the limits of what's possible during emergencies.

Ultimately, the collaborative potential between AI and those on the ground is immense. By bridging the gap between tech innovation and practical application, we empower communities to stand stronger in the face of disaster, moving towards a future where the impact of such events is significantly lessened through foresight and preparedness.

The journey of integrating AI with disaster relief efforts is still evolving. It holds promise not only in enhancing current procedures but also in shaping a new era of smart, anticipatory response frameworks. As we continue to explore these capabilities, smart assistants stand ready to advance the future of disaster preparedness, making the planet a safer place for everyone.

Chapter 24:
The Business of AI Assistants

In recent years, the business landscape for AI assistants has exploded with transformative potential, prompting companies to navigate a rapidly evolving market teeming with opportunities. As digital concierges, these intelligent systems hold the power to revolutionize customer interaction, streamline operations, and unlock new revenue streams. With giants like Amazon, Google, and Apple innovating at breakneck speeds, their cutting-edge developments set the pace for smaller enterprises striving to carve out their niche. Staying ahead means harnessing vast amounts of data and leveraging machine learning algorithms to refine user experience continuously. The race to create value in the AI assistant domain isn't just about technological prowess but also strategic foresight. For businesses, the goal is clear: adapt or risk obsolescence in an age where personalized, responsive technology is not a luxury but a baseline expectation. As companies chart their course, understanding market trends and consumer needs will be vital to capturing the expansive potential that AI assistants offer. The future holds immense promise for those agile enough to seize it.

Market Trends and Opportunities

In the realm of AI assistants, the business landscape is ablaze with transformative trends and untapped opportunities. As technology permeates every facet of our daily lives, AI assistants are positioned not just as technological novelties but as fundamental tools reshaping how we interact with our world. This surge in AI capabilities is fueled by quick advancements in machine learning, natural language processing, and user-centric innovations. Collectively, these developments promise a fertile field of growth and innovation, ripe for exploration by businesses and tech enthusiasts alike.

For starters, market demand for smart assistants is scaling new heights. The accelerating adoption of smart speakers, virtual assistants, and integrated AI platforms exemplifies a trend that extends beyond novelty. These technologies fulfill a pressing societal need for efficiency and convenience, playing a crucial role in managing our increasingly complex personal and professional lives. Corporations and startups are responding by investing heavily in refining these technologies, aiming to offer seamless and intuitive user experiences. As a result, potential emerges for businesses to differentiate themselves via advanced AI solutions specifically tailored to consumer needs.

Integration has become a buzzword in the space, as organizations realize that the true potential of AI assistants lies in their ability to blend effortlessly with other technologies. Interoperability between devices and systems stands as a critical enabler, ensuring users receive a cohesive experience across different platforms. From home automation systems to enterprise software solutions, the harmony achieved through integrated AI platforms is unlocking new avenues for user engagement and productivity. Companies that master this integration art can position themselves at the forefront of the industry's growth trajectory, offering holistic AI ecosystems that captivate users.

Customization, akin to personalization, lays another broad path of opportunity. With AI assistants becoming more ubiquitous, the quest for unique and personalized experiences becomes paramount. Consumers are no longer content with one-size-fits-all services; they crave AI interactions that adapt to their preferences and deliver value-aligned experiences. This movement highlights substantial opportunity for firms to innovate, offering AI solutions customizable at granular levels to cater to diverse user preferences and requirements. These solutions could manifest in the form of tailored AI modules, bespoke command sets, or even personalized interaction styles.

Another distinct trend driving the market is *AI as a Service* (AIaaS). The paradigm shift towards cloud-based AI solutions continues to empower businesses with cutting-edge technological capabilities, without prohibitive infrastructure investments. This progression allows companies of all sizes to harness the power of AI, leveraging it for varied applications from customer support to data analytics. As AI becomes more accessible, companies that focus on developing AIaaS offerings are strategically positioned to capitalize on this democratization, providing invaluable resources to those eager to incorporate AI technologies into their frameworks.

Opportunities frequently present themselves in addressing challenges, and data privacy remains a pivotal concern. AI assistants rely heavily on data, necessitating robust strategies to navigate concerns surrounding data security and user trust. Organizations that prioritize transparent data handling practices and offer users control over their data can distinguish themselves in a crowded market. There is a compelling opportunity for companies that can successfully reassure consumers while delivering powerful AI tools, essentially strengthening their market standing through ethical practices and policies.

Furthermore, regional variations in AI adoption provide another layer of complexity and opportunity. While some regions showcase a voracious appetite for AI innovations, others are more measured, driven by cultural and regulatory considerations. In these nuanced landscapes, there lies a remarkable chance for companies to tailor their approaches, aligning with local expectations, norms, and regulatory environments. This adaptability can be a crucial differentiator, allowing companies to expand their global reach while respecting and integrating cultural distinctions.

The educational sector also represents a burgeoning opportunity. Educational institutions are progressively integrating AI assistants to enhance learning experiences, offering personalized support to students and educators alike. Businesses engaging with the educational sector can tap into this growing demand, developing AI solutions that address unique educational challenges and enhance learning outcomes. As educational methodologies evolve, companies capable of delivering AI-powered educational tools are well placed to revolutionize learning environments.

The healthcare industry beckons as another sector rife with potential. AI assistants in healthcare are facilitating remote monitoring, personalized patient care, and efficient data management. These applications are not only enhancing outcomes but are also optimizing operational efficiencies. Companies adept at navigating healthcare IT constraints and developing compliant solutions stand to make significant contributions to this domain, unlocking opportunities for innovation in patient care delivery and health management.

The entrepreneurial spirit is no stranger to the world of AI assistants, where startups are injecting fresh ideas and disrupting traditional models. Agile and innovative, these enterprises are pioneering new usage cases and generating novel AI applications. Whether it's enhancing user engagement through creative interfaces or developing niche-market products, startups continue to push boundaries and redefine possibilities. By harnessing the energy and creativity of the startup ecosystem, larger tech corporations can foster an environment of collaboration and synergy, driving the industry forward.

Additionally, emerging markets represent uncharted territories ripe for AI assistant expansion. Regions undergoing technological evolution present a vast audience for AI adoption as infrastructures mature and digital literacy rises. Businesses can seize this opportunity by positioning themselves as early movers, establishing a foothold in markets with burgeoning tech ecosystems. The adoption strategies must, however, be thoughtful, respecting local contexts while providing invaluable tools that align with regional advancement goals.

Collectively, these trends and opportunities demonstrate an industry still in its adolescent phase, characterized by rapid evolution and development flywheels. For businesses, the path forward entails navigating these currents with strategic vision, adaptability, and a keen understanding of consumer needs. As AI assistants become ever more entwined with daily life, the businesses that thrive will be those poised at the intersection of technology, user experience, and ethical responsibility. They must harness innovation at every turn, nurturing the human-centric capabilities of their AI solutions while preserving trust and fostering inclusion. Perhaps then, the narrative of AI assistants will evolve not just as tools of convenience but as pivotal companions in our shared journey toward a brighter, more efficient world.

Innovations by Leading Companies

In the dynamic landscape of artificial intelligence, tech giants and startups alike are engaged in a bonafide race to redefine what's possible with AI assistants. These companies are transforming day-to-day activities with innovative solutions designed to enhance convenience, productivity, and user experience. They're developing technologies that weave AI into the very fabric of our lives, crafting a future where AI doesn't just assist but seamlessly integrates. In this space, creativity meets technical prowess, where each company strives to break new ground.

At the forefront, we find companies like Google, Amazon, and Apple, whose AI assistants—Google Assistant, Alexa, and Siri—have become household names. Google Assistant, found in a variety of devices from smartphones to smart speakers, leverages the power of Google's extensive data networks. The result is an assistant that's incredibly adept at natural language processing and instantly provides relevant information from the web. Google's innovation lies in its machine learning algorithms that continually learn and adapt to user preferences, crafting personalized interactions.

Amazon's Alexa has carved a niche in the smart home ecosystem. The range of Alexa-enabled devices illustrates Amazon's commitment to the home automation revolution. Alexa sets itself apart with its compatibility with thousands of smart devices and the ability to control virtually any home gadget. Its Skills platform offers developers a vast playground to create new functionalities, constantly expanding what Alexa can do. This openness has fueled a surge in app development, creating a rich and diverse AI-driven ecosystem.

Apple, on the other hand, harnesses its renowned focus on privacy and security with Siri. Integrated deeply into Apple devices, Siri emphasizes simplicity and ease of use. Apple's meticulous attention to detail ensures Siri works seamlessly with other apps and services within the Apple ecosystem. Recent enhancements have allowed Siri to improve significantly in contextual understanding, making it more responsive and aware of ongoing tasks and user habits.

Microsoft's Cortana, though facing challenges in a market dominated by other giants, isn't to be overlooked. It's been pivoted towards business applications, aiming to enhance productivity tools within Microsoft's suite of products. The integration of Cortana with Office 365 exemplifies Microsoft's strategy to improve workplace efficiency through AI. With features like intelligent calendar management and task reminders, Microsoft intends to transform professional environments into intelligent workspaces.

In the arena of startups, there's a groundswell of innovation as well. These agile companies often tackle niche problems with precision and originality. An example is x.ai, whose virtual assistant automates meeting scheduling with surprising efficiency. Similarly, startups like Viv Labs—acquired by Samsung—focus on groundbreaking technologies that enable AI assistants to handle complex multi-step tasks with evident ease.

Alibaba, pushing boundaries in the Chinese market, has introduced its own AI assistant, AliGenie, integrated into smart home devices. It exemplifies Alibaba's strategy of blending AI with e-commerce, creating a system where users can effortlessly interact with the company's retail services. Alibaba's AI innovations revolve around boosting consumer convenience and enhancing shopping experiences through conversational commerce.

Meanwhile, Facebook has its own ventures, such as M, a personal assistant that's more focused on accurately completing tasks than engaging in social banter. Facebook's unique approach lies in using AI incorporated with human oversight to provide services that require a nuanced understanding of human needs and contexts.

As AI assistants become more ubiquitous, companies are also turning their gaze inward, examining ways these technologies can enhance customer service. IBM Watson, for instance, has found success with customer service bots that offer unparalleled assistance and support. Known for its deep language processing capabilities and analytical skills, Watson assists businesses by automating and optimizing responses to customer inquiries, significantly reducing wait times and improving user satisfaction.

Delivery service companies are another vanguard pushing AI technologies into mainstream business operations. For example, UPS and FedEx utilize machine learning algorithms to optimize delivery routes, anticipate weather disruptions, and reduce fuel consumption. These applications not only improve efficiency but also represent the intersection of AI and environmental stewardship.

In the highly competitive financial sector, firms are increasingly employing AI to handle everything from fraud detection to customer engagement. Goldman Sachs and JPMorgan Chase are leaders in using AI to streamline trading processes and analyze complex market data, ensuring they stay ahead in a rapidly evolving marketplace.

The gaming industry presents a slightly different spin, with companies like NVIDIA using its AI technologies to improve graphics rendering and image processing. This has allowed for more immersive and realistic gaming experiences. Then there's Unity, which integrates AI to improve game design and testing processes, thereby reducing development cycles and costs.

Over in the healthcare industry, substantial advances are being made with AI-powered assistants. IBM Watson Health employs smart assistants to analyze patient data, offering insights for diagnoses and treatment plans. This enhances the decision-making process for healthcare professionals and supports a more personalized approach to patient care.

As we continue exploring these innovations, one can't ignore the persistent challenges that companies face in AI development. Ethical considerations, data privacy, and the digital divide are recurrent themes that require thoughtful navigation. Leading companies are addressing these concerns by pushing for greater transparency in AI operations and emphasizing user consent in data usage policies.

Overall, the rapid evolution of smart assistants is fueled by inventive energies from across the technological spectrum. As these innovations unfold, they push the envelope of what AI can achieve, both in how we integrate these assistants into our lives and the potential they hold to revolutionize the way we interact with technology. Companies that successfully blend technical sophistication with human-centric design will set the bar for the future of AI and its role in society.

Chapter 25:
Developing Your AI Strategy

As we stand at the intersection of technology and daily life, developing a robust AI strategy becomes essential for leveraging smart assistants to their fullest potential. Crafting a long-term AI plan involves understanding your needs and aligning them with the capabilities of AI technology, ensuring that these digital companions are seamlessly woven into the fabric of your daily routine. This isn't just about adopting technology—it's about optimizing its use by setting measurable goals, actively monitoring progress, and refining processes. While the promise of AI is boundless, the challenge lies in sustaining a balance between enrichment and efficiency, ensuring that the deployment is purposeful and adapts alongside the ever-evolving landscape of tech advancements. The opportunity to reimagine productivity and personal growth is right within our grasp, and with a strategic approach, we can transform aspiration into achievement.

Creating a Long-Term AI Plan

Developing a long-term AI plan is crucial in realizing the full potential of smart assistants as they become increasingly integrated into our lives. It involves more than just setting objectives—it's about mapping a journey that aligns with your personal goals and values while staying adaptable to the rapid advancements in technology. This planning process requires foresight, creativity, and a willingness to embrace the changes that artificial intelligence can bring to everyday experiences.

To start, consider what you want your smart assistant to accomplish, not just now but in the future. This involves reflecting on both immediate needs and aspirational goals. Whether you aim to enhance productivity, manage your home more efficiently, or deepen your learning experiences, what you plan today should accommodate growth tomorrow.

A critical aspect of a long-term AI plan is setting realistic yet ambitious goals. These goals should be specific enough to guide daily interactions with your AI but flexible enough to allow for technological advancements and personal growth. Begin by identifying areas in your life where AI can provide maximum benefit. Perhaps it's in managing your calendar, optimizing your work projects, or personalizing your entertainment. Once you've identified these areas, list down tangible outcomes you'd like to achieve using AI.

Think of technology not as a means to an end, but as a continuing partner in your progress. For instance, envision using AI to not just handle mundane tasks but to also inspire and develop creative pursuits. An artist might use AI for brainstorming new ideas or optimizing workflows, turning what were once idle moments into productive sessions of creativity. By framing your goals with both everyday functionality and future ambitions in mind, you'll create an AI plan that's both practical and pioneering.

Another key component in formulating a long-term AI plan is staying informed about emerging technologies. AI is a field known for rapid innovation, with new features and capabilities constantly being introduced. Keeping up-to-date with these developments is essential for modifying and refining your AI strategy. This doesn't mean you need to become an expert overnight; instead, focus on areas that directly impact your objectives. Subscribing to relevant newsletters, attending workshops, or participating in online forums can provide valuable insights.

An informed approach also means understanding the potential challenges that come with AI. Issues like privacy concerns and ethical considerations are critical to discuss—especially when devising a long-term strategy. Make it a point to stay aware of ethical guidelines and privacy policies surrounding AI usage. Having a clear understanding will not only ensure that your use of AI is responsible but will also help in safeguarding your data and maintaining trust in AI systems. These considerations should be woven into your AI strategy to preemptively address potential conflicts.

Balancing technology use with human interaction is another important facet to contemplate. As you integrate AI more deeply into your life, ensure it complements rather than replaces real-world connections. The ideal AI plan fosters productivity and convenience without eroding personal relationships or diminishing emotional intelligence.

Consider how AI can empower not just individual growth but collective well-being. Implement AI tools that support shared objectives, whether it's through collaborative platforms at work or community-focused projects. For example, smart assistants can facilitate more effective communication within teams, assist in co-planning events, or even help in managing local environmental initiatives.

An iterative approach to your AI plan is also essential. Technology and personal goals evolve, and your AI strategy should adapt accordingly. Regularly review your plan—perhaps every six months—and assess what's working and what's not. This ongoing evaluation allows you to make necessary adjustments and incorporate new technologies that align with your original vision but might offer additional benefits you hadn't anticipated.

Measuring success is fundamental to ensuring your AI strategy remains on course. Establish metrics that align with your objectives and regularly analyze your progress against these benchmarks. Keep in mind that success isn't always about achieving more with less time, but also about enhancing the quality of your experiences and the depth of your learning.

At its core, creating a long-term AI plan is about being proactive rather than reactive to the changes AI brings. It's about consistently reimagining the possibilities AI can unlock in your life and making mindful decisions that align with your long-term vision. By committing to a thoughtful, dynamic approach, you ensure that your engagement with AI isn't just an experiment in convenience but a meaningful journey towards enhancing personal and professional lives.

Measuring Success and Optimization

The implementation of an AI strategy is just the starting point. To truly leverage the potential of AI-powered smart assistants, it's essential to focus on measuring success and optimizing performance. This chapter delves into the ways you can evaluate the impact of AI initiatives and continually refine them for better efficiency and productivity. By establishing a robust framework to measure outcomes, you ensure that your efforts not only meet initial expectations but also adapt to evolving needs and challenges.

When we talk about measuring success, it's crucial to set clear, specific goals from the get-go. These are the yardsticks by which you'll measure progress, and they need to be SMART: Specific, Measurable, Achievable, Relevant, and Time-bound. For instance, if you're using a smart assistant to boost workplace productivity, the goal could be to reduce time spent on scheduling meetings by 50% within six months. This gives you a tangible target to aim for and helps in determining whether the AI strategy is fruitful.

After goal setting, the next step is to establish key performance indicators (KPIs) that align with your objectives. KPIs serve as pivotal metrics that reflect the performance of your AI systems. They could include metrics like task completion rates, response accuracy, time savings, user satisfaction, and return on investment. Regularly reviewing these indicators helps identify any gap between the current performance and your goals, thus pinpointing areas that need improvement.

But measuring success isn't just about numbers; qualitative feedback plays a vital role too. User feedback gives rich insights into the user experience beyond what quantitative data can provide. Encouraging feedback through user surveys or interviews can uncover unforeseen issues, highlight unexpected benefits, and provide direction for future development. These insights are invaluable for tailoring the AI to fit user needs more closely.

Once you've gathered the necessary data, optimization comes into play. Optimization is an iterative process that involves tweaking and refining the AI framework based on the insights derived. This can mean updating algorithms to improve efficiency, introducing new features that align with user expectations, or eliminating redundancies that slow down processes. An iterative mindset allows you to make small, regular adjustments that steadily enhance the system's performance.

An essential part of optimizing your AI strategy is keeping an eye on emerging AI technologies and trends. Technological advancements can open doors to new possibilities and dramatically shift how an AI system can be employed. Therefore, staying updated on trends can provide competitive advantages and enable strategic tweaks that align with industry leading practices.

Collaboration is another component that can't be ignored in the optimization process. Engaging with cross-functional teams within your organization can bring diverse perspectives and expertise that benefit the optimization effort. When different departments come together—bringing their unique needs and concerns to the table— your AI system can be better aligned with the organization's overarching goals, ensuring comprehensive, integrated value.

Security and privacy considerations are paramount while optimizing AI systems. Ensuring data privacy and maintaining robust security mechanisms not only safeguard user information but also enhance trust and reliance on AI assistants. Optimization efforts must strike a balance between improved capabilities and stringent privacy controls, protecting sensitive data while achieving desired outcomes.

Moreover, scalability should be factored into your optimization initiatives. As the technology landscape evolves or your organization grows, the AI strategy should be capable of expanding without compromising performance or user experience. This may involve optimizing algorithms for scalability, leveraging cloud computing, or employing modular designs that allow for seamless integration of new features and adaptations.

Financial considerations should be part of your success and optimization plan too. Regular cost-benefit analyses will ensure that the AI system is economically viable. Consider the costs related to implementing new technologies, maintaining existing systems, and potential savings from reduced overhead or improved efficiencies. Balancing these financial aspects with performance metrics provides a comprehensive view of an AI strategy's success.

Finally, it's important to cultivate a culture of continuous improvement within your environment. Encouraging a mindset that embraces change and innovation can sustain long-term success and optimization. By fostering this culture, employees will be more inclined to engage with AI systems and provide feedback, ensuring that your AI strategy remains agile and adaptable to new challenges and opportunities.

In conclusion, measuring the success of AI-powered smart assistants is a multifaceted endeavor that extends beyond initial implementation. By setting clear goals, using quantitative and qualitative metrics, and adopting an iterative optimization approach, businesses and individuals can maximize the potential of AI technologies. This dynamic process not only enhances productivity and efficiency but ultimately aligns AI capabilities with the ever-changing demands of the modern world.

Conclusion

The journey of integrating AI-powered smart assistants into our lives has just begun, yet the impact is already undeniable. These digital companions have evolved from mere curiosities to essential tools that aid us in managing our daily routines, enhancing our productivity, and even protecting our time with family and friends. As we reflect on the potential of smart assistants, it's clear that their promises are vast and varied, interweaving themselves into the very fabric of contemporary living.

Our exploration began with understanding what smart assistants are and how they've transcended their initial limitations. From simple voice-activated tasks to complex data-driven decision-making, these assistants have become far more sophisticated and intuitive. They've infiltrated almost every aspect of our day-to-day activities, reshaping how we work, learn, relax, and connect with others.

Incorporating smart assistants into daily activities provides a unique opportunity to automate mundane tasks, freeing us to focus on more meaningful engagements. The convenience of automating our schedules, setting reminders, and even managing household appliances through voice commands exemplifies this transformation. It's not just about efficiency, but about empowering individuals to lead more fulfilling lives.

Nevertheless, the deployment of smart assistants brings along challenges that call for careful consideration. Privacy concerns remain a sticking point. Understanding data security and being vigilant about digital footprints have become as crucial as the technology itself. Users must navigate these waters by employing best practices to safeguard their information without sacrificing convenience.

Furthermore, as AI technology advances, the potential for personalization and enhancement grows. In education, health, finance, and beyond, AI offers tools for customization that were once considered futuristic. By tailoring experiences to individual needs, smart assistants can foster an environment of supportive, personalized growth that aligns with personal goals, whether it be honing skills, achieving fitness targets, or managing investments more wisely.

Yet, with this great power comes the responsibility of maintaining a balanced relationship with technology. There's a fine line between dependency and benefit, and it's imperative to draw this line wisely. Promoting meaningful real-world connections while utilizing digital tools for convenience requires a harmonious balance, encouraging a healthy interplay between digital interaction and human connection.

In looking to the future, the landscape of smart assistants is full of possibilities. As they become more deeply ingrained into various sectors, advancements in AI capabilities promise to further blur the lines between technology and human functionality. The work environment is expected to shift dramatically, with AI assisting in core tasks but also requiring a recalibration of traditional roles.

Finally, as we anticipate future developments, it is vital to engage thoughtfully with the cultural and ethical considerations that accompany this technology. AI's impact is global, influencing diverse cultures in varied ways. Equally, ethical paradigms around privacy, data ownership, and equity must be continually reexamined to ensure AI's safe and fair use.

In conclusion, smart assistants hold a transformative power that is only beginning to be realized. They challenge us to rethink how we interact with technology and its potential to enrich our lives. As we continue to embrace these advancements, the goal should be to leverage them in a manner that amplifies human capacity, fosters innovation, and propels us towards a future where technology and humanity coexist in a beneficial synergy. The journey with smart assistants is not a finite destination but an ongoing voyage that promises both challenges and unforeseen enhancements in every aspect of our lives. As we tread this path, let us do so mindfully, harnessing the full spectrum of opportunities this new digital age has to offer.

Appendix A:
Appendix

As we explore the vast realm of artificial intelligence and smart assistants, this appendix serves as a foundation, offering additional insights and resources to enrich your understanding and application of these transformative tools. With the tech-laden world evolving at a breakneck speed, staying ahead means diving deeper into the nuanced aspects of AI that this section unravels.

Smart assistants are becoming indispensable allies, seamlessly weaving themselves into the fabric of daily life and revolutionizing how we approach tasks. From managing personal budgets to ensuring home security, their capabilities continue to expand, driven by advancements in machine learning and data processing.

Moreover, beyond the convenience and efficiency they offer, there lies a complex web of ethical, cultural, and legal considerations that this appendix helps navigate. Recognizing these aspects is crucial for not only leveraging AI to its fullest potential but also ensuring its responsible application in various contexts.

For those looking to delve deeper, a multitude of online forums, scholarly articles, and books are available. Engaging with these resources will not only enhance your knowledge but also keep you updated on the latest trends and innovations shaping the future of smart assistants.

Whether you're refining your AI strategy or merely seeking to better understand your device's capabilities, this appendix is designed to guide you in your journey. Embrace the opportunity to harness the power of AI with an informed and balanced perspective, ensuring its applications elevate both personal productivity and societal advancement.